PROTEIN BIOSYNTHESIS IN BACTERIAL SYSTEMS

EDITED BY

Jerold A. Last and Allen I. Laskin

National Academy of Sciences
Washington, D.C.

ESSO Research and Engineering
Company
Linden, New Jersey

MARCEL DEKKER, INC.   New York   1971

Copyright © 1971 by MARCEL DEKKER, INC.

ALL RIGHTS RESERVED

No part of this work may be reproduced or utilized in any form or by any means, electronic or mechanical, including *Xeroxing, photocopying, microfilm, and recording,* or by any information storage and retrieval system, without permission in writing from the publisher.

MARCEL DEKKER, INC.
95 Madison Avenue, New York, New York 10016

LIBRARY OF CONGRESS CATALOG CARD NUMBER: 78-160517
ISBN NO.: 0-8247-1396-6

PRINTED IN THE UNITED STATES OF AMERICA

PROTEIN BIOSYNTHESIS
IN BACTERIAL SYSTEMS

METHODS IN MOLECULAR BIOLOGY

Edited by

ALLEN I. LASKIN
ESSO Research and Engineering Company
Linden, New Jersey

JEROLD A. LAST
National Academy of Sciences
Washington, D.C.

VOLUME 1: Protein Biosynthesis in Bacterial Systems, *edited by Jerold A. Last and Allen I. Laskin*

VOLUME 2: Protein Biosynthesis in Nonbacterial Systems, *edited by Jerold A. Last and Allen I. Laskin*

VOLUMES IN PREPARATION

VOLUME 3: Nucleic Acids, *edited by Allen I. Laskin and Jerold A. Last*

VOLUME 4: Subcellular Particles, Structures, and Organelles, *edited by Allen I. Laskin and Jerold A. Last*

PREFACE

Every laboratory research worker has experienced the many frustrations associated with attempting to duplicate a procedure described in the conventional scientific literature. Rarely does a paper describe the experimental methods in a manner that allows one readily to reproduce them in his own laboratory. This problem has resulted in the appearance of a variety of books, manuals, etc., on "methods," "techniques," and "procedures." In many instances, however, the descriptions found in such books may be no easier to follow than those in the original literature; they may be "buried" in a volume of unrelated techniques, and often may be in a book with a rather high cost.

Methods in Molecular Biology represents an attempt to provide small, relatively inexpensive, topically organized volumes, which might be particularly beneficial to new workers in a field, to graduate students beginning a problem, to new technicians, etc.

The authors were asked to write descriptions of the methods used in a particular area as critically as possible, and whenever appropriate, to discuss such things as: _why_ a particular approach was taken, _why_ a particular reagent was used, what alternatives are feasible and acceptable, and what to do "if things go wrong," etc.

In treating the subject of *in vitro* protein biosynthesis in bacteria, this volume begins with a description of some of the "crude" (S-30) cell-free systems, first from *Escherichia coli*, and then examples of other systems: a gram-positive (*Bacillus subtilis*) and a specialized (halophilic) system. The case in which the *in vitro* synthesis is directed by DNA is followed by chapters concerned with various purified components of the protein-synthesizing machinery—ribosomes; initiation, elongation, and termination factors; aminoacyl-tRNA synthetases; tRNA; phage messenger RNA; ribosomal RNA. Volume 2 of the series will deal with systems other than bacteria.

Washington, D. C.   Jerold A. Last
Linden, New Jersey   Allen I. Laskin
August 1971

CONTRIBUTORS TO THIS VOLUME

S. T. BAYLEY, Department of Biology, McMaster University, Hamilton, Ontario, Canada

ROY H. DOI, Department of Biochemistry and Biophysics, University of California, Davis, California

JACK GOLDSTEIN, New York Blood Center and Cornell University Medical College, New York, New York

JULIAN GORDON, The Rockefeller University, New York, New York

MASAKI HAYASHI, Department of Biology, University of California, San Diego, La Jolla, California

RAYMOND KAEMPFER, The Biological Laboratories, Harvard University, Cambridge, Massachusetts

DANIEL KOLAKOFSKY, Institut für Molekularbiologie, Universität Zürich, Zürich, Switzerland

JEROLD A LAST,* The Rockefeller University, New York, New York

JUAN MODOLELL, Instituto de Biología Celular, C.S.I.C., Velazquez, Madrid, Spain

KARL H. MUENCH, Division of Genetic Medicine, Departments of Medicine and Biochemistry, University of Miami School of Medicine, Miami, Florida

EDWARD M. SCOLNICK, Viral Lymphoma and Leukemia Branch, National Cancer Institute, Bethesda, Maryland

FREDERICK VARRICCHIO, Department of Internal Medicine, Yale University Medical School, New Haven, Connecticut

* Present address: National Academy of Sciences, Washington, D.C.

CONTENTS

Preface iii

Contributors to This Volume v

Chapter 1. THE S-30 SYSTEM FROM Escherichia coli
Juan Modolell

I. Introduction 2
II. Bacterial Strains 3
III. Growth of Cells 4
IV. Preparation of S-30 Extracts 6
V. Preparation of S-100 Supernatant and Ribosomes . . . 10
VI. Polypeptide Synthesis 12
VII. Analysis of Polypeptide-Synthesizing Systems in Sucrose Density Gradients 32
VIII. Extraction of Polysomes from E. coli Cells 48
References 58

Chapter 2. Bacillus Subtilis PROTEIN-SYNTHESIZING SYSTEM
Roy H. Doi

I. Introduction 68
II. Growth of B. subtilis Cells and Spores 70
III. Preparation of Cell-Free Extracts for Protein Synthesis 75
IV. Assay Systems for Protein Synthesis 76
V. Conclusions 83
References 84

Chapter 3. PROTEIN SYNTHESIS SYSTEMS FROM HALOPHILIC BACTERIA

S. T. Bayley

I.	Introduction	89
II.	Growth and Harvesting of Bacteria	91
III.	Homogenization of Cells and Preparation of S-60 Extract	96
IV.	Assay of S-60 Extract	98
V.	Preparation of Ribosomes Free of mRNA and of S-150 Extracts	101
VI.	Incorporation with Synthetic mRNAs	103
VII.	Preparation of Aminoacyl-tRNA Synthetases and tRNA	104
VIII.	Measurement of Aminoacyl-tRNA Formation	107
	References	109

Chapter 4. DNA-DEPENDENT, RNA-DIRECTED PROTEIN SYNTHESIS

Masaki Hayashi

I.	Introduction	111
II.	Materials	114
III.	Procedure	115
IV.	Assay for RNA and Protein Synthesis in the Coupled System	116
V.	Application of the System	117
	References	118

Chapter 5. RIBOSOMAL SUBUNIT EXCHANGE AND DENSITY GRADIENT CENTRIFUGATION

Raymond Kaempfer

I.	Introduction	121
II.	Principle of the Technique	122
III.	Procedure	124
IV.	Applications	144
V.	Other Methods	146
	References	147

Chapter 6. CHAIN INITIATION FACTORS FROM Escherichia coli

Jerold A. Last

I.	Introduction	151
II.	Sources	154
III.	Preparation of Crude Factors	154
IV.	Assays	169
	References	174

Chapter 7. CHAIN ELONGATION FACTORS

Julian Gordon

I.	Introduction	177
II.	General Features of the Preparation	179
III.	Growth Conditions and Preparation of Extracts	181
IV.	Separation of T and G Factors	188
V.	Separation of T_U and T_S	192
VI.	Recovery of the Ribosomes	192
VII.	Assays	194
	References	197

Chapter 8. PREPARATION OF POLYPEPTIDE TERMINATION FACTORS FROM Escherichia coli

Edward M. Scolnick

I. Introduction 201
II. Preparation of Materials 203
III. f[^{3}H]Met-tRNA-AUG-Ribosome Complex 206
IV. Release Assay. 207
V. Preparation of R1 and R2 208
References 211

Chapter 9. PREPARATION OF AMINOACYL-tRNA SYNTHETASES FROM Escherichia coli

Karl H. Muench

I. Introduction 214
II. Methods of Assay 216
III. General Approach to Purification of Synthetases. . . 221
References 230

Chapter 10. TRANSFER RNA

Jack Goldstein

I. Introduction 235
II. Isolation of Crude tRNA. 238
III. Further Purification 250
IV. Fractionation Methods. 257
References 262

Chapter 11. PREPARATION OF COLIPHAGE RNA

Daniel Kolakofsky

Contents   xi

I.   Introduction 267
II.   Coliphage Growth 268
III.   Coliphage Purification 270
IV.   Preparation of Coliphage RNA 273
       References 276

Chapter 12.   RIBOSOMAL RNA

Frederick Varricchio

I.   Introduction 279
II.   Bacterial Strains and Culture Conditions 280
III.   Preparation of Cell Extracts 282
IV.   Preparation and Lysis of E. coli Spheroplasts. . . . 283
V.   Preparation of Ribosomes 284
VI.   Extraction of RNA. 288
VII.   Separation of RNA. 290
VIII. Polyacrylamide Gel Separation of RNA 291
IX.   Characterization of RNA. 300
       References 312

Author Index. 317
Subject Index   329

PROTEIN BIOSYNTHESIS
IN BACTERIAL SYSTEMS

Chapter 1

THE S-30 SYSTEM FROM Escherichia coli

Juan Modolell

Instituto de Biología Celular
C.S.I.C.
Velazquez, 114
Madrid,
Spain

I.	INTRODUCTION	2
II.	BACTERIAL STRAINS.	3
III.	GROWTH OF CELLS.	4
IV.	PREPARATION OF S-30 EXTRACTS	6
V.	PREPARATION OF S-100 SUPERNATANT AND RIBOSOMES . . .	10
VI.	POLYPEPTIDE SYNTHESIS.	12
	A. Synthesis with Natural Messenger	18
	B. Synthesis with **Synthetic** Polynucleotides	30
VII.	ANALYSIS OF POLYPEPTIDE-SYNTHESIZING SYSTEMS IN SUCROSE DENSITY GRADIENTS.	32
	A. Preparation of Sucrose Density Gradients	33
	B. Zonal Centrifugation	35
	C. Analysis of Gradients.	36
	D. Interpretation of Gradient Profiles.	40

Copyright © 1971 by Marcel Dekker, Inc. No part of this work may be reproduced or utilized in any form or by any means, electronic or mechanical, including xerography, photocopying, microfilm, and recording, or by any information storage and retrieval system, without the written permission of the publisher.

VIII. EXTRACTION OF POLYSOMES FROM E. coli CELLS 48

    A. Cell Disruption by Freeze-Thaw-Lysozyme Treatment. 49

    B. Cell Disruption by EDTA-Lysozyme Treatment . . . 54

    C. Polysome Extraction by Other Procedures. 57

REFERENCES 58

I. INTRODUCTION

The S-30 system from Escherichi coli was first described in 1961 by Nirenberg and Matthaei [1]. It consists of a crude DNase-treated extract of E. coli, freed of cells and cell debris by centrifugation at 30,000 X g, containing most of the components necessary for polypeptide synthesis (ribosomes; tRNA; amino acid-activating enzymes; initiation, elongation, and termination factors; and so on) plus many other more-or-less ill-defined materials (S-30 extract). The extract is usually preincubated to deplete its content of endogenous messenger (iS-30 extract). It is supplemented with amino acids, ATP, an ATP-regenerating system, GTP, and a synthetic or natural mRNA. With adequate concentrations of Mg^{2+} and K^+ or NH_4^+ ions, such a system can carry out extensive incorporation of amino acids into polypeptides; it is capable of synthesizing complete specific proteins [2-5] and enzymes [6].

Despite the advent of more purified and better defined systems for amino acid incorporation, the S-30 system is still preferred in many experiments because of its simplicity of

1. The S-30 System from Escherichia coli

preparation, stability, and high activity. Moreover, when directed with natural messenger, it fairly reproduces most of the steps of protein synthesis in the whole cell: physiological initiation with N-formylmethionyl-tRNA (fMet-tRNA), ribosomal subunits, and initiation factors; elongation with extensive formation of polysomes; physiological termination with release of the completed polypeptide chain; and recycling of ribosomes for new rounds of synthesis. Thus it affords an in vitro polypeptide-synthesizing system suitable for the study of any of these steps.

In this chapter procedures are given for growing cells, preparing the S-30 extract, setting up the amino acid-incorporating system, determining the amount of synthesized product, and analyzing the system in sucrose density gradients. As a complement to the S-30 system, the preparation of cell-free extracts under conditions mild enough to preserve the cell polysomes is also discussed.

II. BACTERIAL STRAINS

Escherichia coli B and K12 have been the most commonly used strains for the preparation of S-30 extracts. However, in principle, any strain should be suitable. Strains deficient in RNase I, such as MRE 600 [7] or Q13 (also deficient in polynucleotide phosphorylase [8]), are becoming increasingly popular and their use is recommended. In our experience, comparing strains S26 (an RNase I$^+$, K12 strain [9]) and MRE 600, MRE 600 formed polysomes in an S-30 system directed with R17 RNA that were more stable under prolonged incubations than those formed by S26. Still, the activity of both S-30 systems, measured as the total number of amino acids incorporated per ribosome, was about the same.

III. GROWTH OF CELLS

Normally, bacteria are grown at 37°C in a rich medium such as nutrient broth or L medium [10]: 10 g Bacto-tryptone, 1 g yeast extract (both Difco products), and 5 g NaCl in 1 liter of water, to which glucose is added (to 0.2%) after autoclaving (4 ml of 50% glucose per liter of medium). Growth is conducted in large Erlenmeyer flasks (up to 1.5 liters of medium in a 6-liter flask) on an appropriate shaking incubator, or in a large carboy or fermentor under forced vigorous aeration. Air should be filtered by passing it through flasks filled with cotton wool and injected into the medium through three or four tubes terminated in fritted-glass plaques to disperse it into fine bubbles. Since excessive foaming may occur under these conditions, the use of an appropriate sterile antiform is indicated (minimal amounts should be used).

To start growth the preheated medium is inoculated with 1/100 vol of a fresh stationary phase culture (a few-days-old culture can also be used if it has been kept refrigerated). The cells are grown to midlog phase, that is, to approximately 1×10^9 cells per milliter (equivalent to an A_{490} of 0.45 in a

1. The S-30 System from *Escherichia coli*

Lumetron colorimeter, or 140 Klett units at 490 nm in a Klett-Summerson colorimeter). Growth is terminated by rapidly cooling the cells; pour the culture onto crushed ice or add about one-fifth the weight of the culture of ice to the flask and agitate. Cells are collected by centrifugation in the cold at 10,000 X g for 10 min or in a continuous-flow centrifuge (Sharples). The yield is approximately 2 g of packed cells per liter of culture. Lodish [11] recently reported that cell yields as high as 13-20 g per liter can be obtained by using a more concentrated medium and letting growth proceed up to saturation. Extracts from these cells also appear to be very active.

Immediately after they are harvested, the cells are washed at 0°C by resuspension in approximately 25 ml of standard buffer [10 mM tris-HCl (pH 7.8), 60 mM NH_4Cl, 10 mM magnesium acetate, 6 mM 2-mercaptoethanol; see Section IV] per liter of original culture, and collected by centrifugation at 20,000 X g for 10 min. Unless the cells are going to be used immediately, they should be quickly frozen. This can be conveniently done by immersing the centrifuge tube (stainless steel or polypropylene) in a dry ice-acetone mixture. The frozen pellet can be quickly weighed and stored wrapped in plastic or aluminum foil. Cells can be stored at -70°C or in liquid nitrogen for weeks (and possibly months) and still yield extracts as active as those prepared from fresh cells. However, storage at -20°C for a substantial period of time results in extracts with decreased activity when assayed with natural messenger. Temperature of storage may be less critical if the extracts are used with synthetic messengers such as poly U.

IV. PREPARATION OF S-30 EXTRACTS

Grinding with alumina, under the conditions described by Nirenberg and Matthaei in 1961 [1], is still the method most widely used to break cells. Many variants of their procedure have been published [2-4, 11-14]. The one described here has been used repeatedly in our laboratory and consistently yields extracts of satisfactory activity and stability. Other alternative methods, such as the French pressure cell, can also be employed successfully [12-15]. Sonication, however, is not recommended. Milder methods, which preserve the polysomes of the intact cell, and use of hydrolytic enzymes to attack the cell wall and detergents to lyse cells, are discussed later (Section VIII). Generally, they are not adequate for preparing amino acid-incorporating systems directed by exogenous messengers.

All the operations are done between 0 and 4°C. Grinding with alumina is usually done in a cold room with a chilled mortar and pestle. Alternatively, it can be done in the laboratory if the mortar is placed in a larger container and surrounded with well-packed crushed ice (this alternative should not be followed when small mortars are used since it is difficult to avoid some ice chips falling into the mixture during grinding).

1. The S-30 System from *Escherichia* coli

The amount of cells ground in most cases is between 2 and 30 g, depending on the requirements of the experiment. As a guideline, consider that 3 g of cells yield about 5 ml of S-30 extract with approximately 12 mg/ml of ribosomes, an amount suitable for many amino acid incorporation experiments. The mortar and pestle should be made of unglazed porcelain. An 8-cm diameter mortar is suitable for smaller amounts, while one of 15-20 cm is appropriate for larger amounts. A pestle with a fairly flat ginding end is more effective than one with a near-spherical end.

If frozen cells are used, the pellet or pellets should be cracked into small pieces in the mortar. The cells are then ground with 1-1/2 to 2 times their weight of alumina (Norton levigated alumina, or Alcoa A-301) until the mixture is perfectly homogeneous, has a chewing-gum consistency, and emits popping sounds. The correct end point is not easily described and is dictated by experience; undergrinding results in diminished ribosome yields, while overgrinding may damage some components of the extract, thus diminishing its activity. The paste is then extracted with 2 vol of standard buffer per gram of ground cells. It is helpful to add the buffer in several portions, reserving a few of them to rinse the mortar after the mixture has been removed. The rinses are mixed and well homogenized with the rest of the extract. Alumina, whole cells, and debris are then sedimented by centrifugation at 20,000 X g for 15 min. The turbid supernatant (it should be quite viscous at this stage because of its content of DNA) is carefully aspirated with a Pasteur pipet. Electrophoretically pure DNase is added to a concentration of 3 µg/ml (DNase preparations of lower purity should not be used since they usually contain small amounts of RNase, which is very inhibitory to the system). After mixing well (taking care not to produce excessive foam) the extract is left standing at 0°C for a few minutes. Viscosity should decrease, and the extract is then clarified by centrifugation at 30,000 X g for 35 min, dialyzed against 500 vol of standard buffer for 4 or 5 hr, and again clarified by centrifugation at 30,000 X g for 20 min. Care should be taken to use narrow dialysis tubing to increase the efficiency of the dialysis. Many investigators use longer dialysis times [1-3,11,12], but in our laboratory this results in diminished activity of the extract. The dialyzed and clarified S-30 extract should have an amber-yellowish color without apparent turbidity. Its A_{260} is determined on a 1-to-500 dilution. The measured A_{260} times 500, multiplied by the factor 0.04, gives the approximate content of ribosomes in milligrams per milliliter. Since S-30 extract contains almost exclusively 70 S ribosomes and native 50 and 30 S subunits, more precise determinations can be performed by separating ribosomes and and subunits from the lower-molecular-weight components of S-30 extract in a sucrose density gradient (see Section VII).

1. The S-30 System from Escherichia coli

The fractions containing ribosomes and subunits are pooled, the A_{260} is measured, and the content of ribosomes plus subunits is calculated by multiplying by the factor 0.06.

Since S-30 extract should never be frozen and thawed more than once, it is divided into small portions, quickly frozen in a dry ice-acetone mixture, and stored at -70°C or in liquid nitrogen. Under these conditions S-30 extract is usually stable for several months. Storage at -20°C results in rapid loss of activity.

Remarks. If disruption of cells with a French press is preferred, frozen cells should be suspended in twice their weight of standard buffer and disrupted in a prechilled cell at 18,000 psi. Additional 2-mercaptoethanol (6 μmoles/ml) is added to the broken cell suspension to compensate for its oxidation [15]. From that point on, the procedure is the same as that for alumina-ground cells.

The standard buffer described by Nirenberg and Matthaei [1] contained KCl instead of NH_4Cl. Either of the two salts can be used, according to the requirements of the experiment. However, NH_4Cl is more commonly used since it seems to yield extracts with somewhat greater activity. Since diluted 2-mercaptoethanol is easily oxidized, the standard buffer should be prepared fresh daily. We find it convenient to maintain a stock of 600 mM NH_4Cl and 100 mM tris-HCl adjusted to pH 7.8, which is diluted 10 times and supplemented with magnesium acetate and 2-mercaptoethanol just before use.

If preferred, DNase can be added, to the same final concentration as above, while the cells are being ground.

Commercial dialysis tubing is normally quite dirty. It should be well cleaned before use. This can be accomplished by boiling it in water, in 0.1 M $NaHCO_3$, twice more in water, once in 1 mM EDTA (30 min each time), and finally washing it with water. It should be stored in 1 mM EDTA.

V. PREPARATION OF S-100 SUPERNATANT AND RIBOSOMES

In some experiments it may be desirable to separate the ribosomes from the remaining soluble components of an S-30 extract. This is normally accomplished by centrifuging undialyzed S-30 extract at 105,000 X g for 3 hr or 160,000 X g for 2 hr. If required, dialyzed S-30 extract or preincubated extract (which has a depleted content of endogenous messenger; see Section VI,A,2) can also be used. The top one-half or two-thirds of the supernatant is carefully aspirated, without disturbing the bottom layers of the fluid, and it is dialyzed, clarified, and stored as described for S-30 extract. This constitutes the S-100 fraction [1] which contains tRNA, the soluble factors G and T (see Chapter 7), termination factors (see Chapter 8), and aminoacyl-tRNA synthetases [16,17] (see Chapter 9), in addition to many other more-or-less defined components. It also contains residual ribosomes and subunits (mostly 30 S) which can be eliminated by a second centrifugation.

1. The S-30 System from Escherichia coli

The S-100 fraction is used as the starting material to prepare any of these components in purified form (see other chapters of this book). It is also used to supplement ribosomes in amino acid-incorporating systems and to enhance the activity of S-30 extracts (see Section VI).

The ribosomal pellet is normally packed, transparent, and a light-amber color. It is overlayered by a brown, gelatinous, loosely packed material that usually slides to the bottom of the tube if centrifugation has taken place in an angle rotor. This material is rich in native 30 S ribosomal subunits [18] and can be used for the preparation of initiation factors (see Chapter 6); it is normally discarded, together with the overlaying supernatant, if the ribosomes in the pellet are to be purified or used without further purification to complement S-100 fractions in amino acid incorporation experiments. In the latter case, the tube walls and pellet should be rinsed with standard buffer, and the ribosomal pellet worked gently into solution in standard buffer with the help of a glass stirring rod or Teflon pestle. After all the transparent ribosomal material is dissolved, the solution is clarified by centrifugation at 30,000 X g for 20 min and the supernatant is carefully aspirated. Its ribosome content can be estimated by its A_{260} (A_{260} X 0.06 = milligrams per milliliter of ribosomes). If gross contamination with supernatant is to be avoided, the ribosomes can be washed by resedimentation

in an ultracentrifuge. In any case, crude ribosomes are
contaminated with degradative enzymes and consequently are more
unstable than purified ribosomes. Thus the ribosomal
suspension should be divided into small portions, frozen, and
stored similarly to S-30 extracts. It may be advisable to
replace 2-mercaptoethanol by 1 mM dithiothreitol (DTT, Cleland's
reagent) in the final resuspension buffer, since this compound
is a more stable reducing agent.

VI. POLYPEPTIDE SYNTHESIS

S-30 extracts can be made to incorporate amino acids into
polypeptides in response to either endogenous or exogenous
mRNA. Systems with DNase-treated S-30 extracts, directed by
endogenous messenger, have low activity in terms of the number
of amino acids incorporated per ribosome. Table I shows that
the average incorporation in two experiments, with two different
extracts, was only 0.52 molecules of [^{14}C]valine per ribosome,
while the incorporation with added messenger (MS2 phage RNA)
was 26.4 molecules. If we consider that E. coli total protein
contains 6% valine [19], and that the main product of the
system directed by the viral messenger is the coat protein of
the phage [2-4], which contains 11% valine [20], these
activities represent the incorporation of 10 and 240 amino
acids per ribosome with endogenous and exogenous mRNA,
respectively. The low activity is a consequence of fragmentation of cell polysomes during alumina grinding, which leaves

1. The S-30 System from Escherichia coli

TABLE 1

Activity of S-30 Extracts

with

Endogenous and Exogenous (Viral) Messenger[a]

Messenger	Preincubation time (min)	Molecules [^{14}C]valine incorporated per ribosome
Experiment 1		
MS2 RNA	7	26.4
Endogenous	0	0.75
Background	7	0.18
Experiment 2		
Endogenous	0	0.30
Background	7	0.12

[a]Reactions, as described in the text, were allowed to proceed for 20 min. In the measurement of background incorporation in experiment 2, the supplementary supernatant and the extract were preincubated together; omission of the supernatant did not significantly affect the results.

the resulting single ribosomes with short pieces of mRNA attached to them. Under these conditions the ribosomes seem to read only the residual portions of untranslated messenger [21]. Thus the incorporation activity mostly represents limited peptide chain elongation [21-23]. If we omit DNase treatment of the extract, the activity of the system increases [21,24,25] since in the presence of an ATP-regenerating system ribonucleoside triphosphates are maintained at a concentration high enough to allow some de novo synthesis of mRNA. Synthesis then is no longer limited to chain elongation, as extensive chain initiation also takes place [21]. Synthesis with endogenous messenger represents "bona fide" polypeptide synthesis as shown by its requirements [24,25] and its sensitivity to ribosomal inhibitors such as streptomycin [23], chloramphenicol, and puromycin [24,25]. For some purposes, however, the alternative systems described in Section VIII may be more appropriate for studying polypeptide synthesis with endogenous messenger.

Since 1961, when Nirenberg and Matthaei discovered that poly U can act as messenger and direct the synthesis of polyphenylalanine, systems with exogenous messengers have become the most widely used. The most common messengers are the homopolynucleotides poly U, poly C, poly A, and poly I (used instead of poly G, which is inactive as messenger because of its secondary structure); some random copolymers of several ribonucleotides, especially poly AUG; the genome of the small RNA bacteriophages R17, MS2, f2, Qβ, and so on; and RNA extracted from E. coli cells infected with virulent DNA phages such as T4. The messenger of choice depends on the nature of the particular experiment to be performed. Natural messengers,

1. The S-30 System from *Escherichia coli*

especially RNA of the small bacteriophages, have become increasingly popular in the last few years, since they offer several distinct advantages over the synthetic polynucleotides. Since they posses initiating and terminating codons, they permit physiological initiation with fMet-tRNA, ribosomal subunits, and initiation factors, as well as physiological termination with termination factors and release of the synthesized polypeptide from the ribosome after its completion (reviewed in Ref. 17). In contrast, with synthetic homopolynucleotides which lack these signals initiation proceeds without factors and fMet-tRNA and requires an artificially high concentration of Mg^2 to promote formation of messenger-ribosome-aminoacyl-tRNA complex [26-30]. The termination mechanism is ineffective and the synthesized polypeptide remains attached to the ribosome [29-31]. Moreover, the high Mg^2 concentration required for activity may impair the fidelity of translation, inducing the incorporation of amino acids not coded by the polynucleotide (misreading [32,33]). However, the fidelity of translation achieved by the S-30 system directed with exogenous natural messenger is exemplified by the synthesis of active lysozyme (with T4 messenger [6]) and of the coat protein [2,4], the specific RNA synthetase [4], and the A protein or maturation protein [5], which are coded by the genome of the small RNA phages. The fidelity of translation is also demonstrated in the in vitro system by early termination at a nonsense mutation and suppression with appropriate genetic suppressors [14,34a]. Moreover, with natural messenger, mechanisms of translational control are operative in vitro: for example, the frequency of initiation at every one of the three cistrons of the RNA phage messenger is regulated and seems to determine the relative amounts synthesized of each of the three proteins (about 1 maturation protein per 6 synthetase subunits and 20 coat proteins [35]).

Another distinct advantage of a system with natural messenger is its more physiological and complete response to antibiotic inhibitors of protein synthesis. Streptomycin is a typical example. This antibiotic completely inhibits polypeptide synthesis in in vitro systems with phage messenger [22,23,36], as it does in E. coli cells. However, it only partly inhibits synthesis with poly U [37-39]. Moreover, even this apparent partial inhibition may not be real, since streptomycin induces misreading which, under some conditions, more than compensates for the inhibition of correct reading [33]. Thus the in vitro system directed with poly U inadequately reproduces the interaction of streptomycin and ribosomes in the cell. The high Mg^2 requirement and the artificial initiation, which may lead to a different interaction of the ribosome with the messenger, are likely explanations for the abnormal response. Other antibiotics, such as chloramphenicol, erythromycin, and spectinomycin, which do not induce misreading and have diverse

1. The S-30 System from *Escherichia coli*

modes of action different from that of streptomycin (reviewed in Ref. 40), are also poor inhibitors in the poly-U system [36,41] and good inhibitors in natural messenger-directed systems [24,36,and our unpublished observations]. With other homopolynucleotides the effect of these antibiotics varies from poor to complete inhibition (specific details for these and other antibiotics can be found in Refs. 41 and 42). In contrast, tetracycline and streptogramin A inhibit as strongly with natural or synthetic messengers (reviewed in Refs. 43 and 44). This variability of responses makes it advisable to compare, whenever possible, the results obtained with a particular messenger and inhibitor with the response to other messengers, especially the natural ones.

Still another advantage of a system directed by RNA phage messenger is its high degree of activity, which can lead to extensive formation of polysomes in vitro (Section VII). This system provides a simple means for studying the behavior of polysomes and for estimating the number of active ribosomes in the system at any time (Section VII,D,1).

These facts may explain the increasing use of S-30 systems directed by natural messengers. Still, systems directed with synthetic messengers, especially homopolynucleotides, have advantages of their own and can provide a quite satisfactory model for protein synthesis. Thus these systems have simpler requirements, since initiation factors, fMet-tRNA, and termination factors are not needed. They focus mainly on the chain elongation cycle and, as a consequence, studies with purified systems directed with synthetic messengers have helped greatly in understanding the workings of this cycle [16]. The homogeneity of the synthesized product may also be quite valuable in simplifying analytical procedures; for example, it is easy to assay for the puromycin reaction in crude systems directed with poly U by the cresol method [45], which is based on the solubility of the released product, polyphenylalanyl puromycin, in this solvent; also, the di- and polylysine peptides synthesized in the poly-A system may be separated by chromatography up to the 14-residue-long tetradecamer peptide [46]. The simpler requirements of the system provide an additional advantage; it is easier to prepare extracts with relatively good activity. The extracts can be handled more roughly or prepared from older cells, the messengers are commercially available, partial hydrolysis by RNAse may not be as critical, and so on.

A. Synthesis with Natural Messenger

Table 2 shows the composition of a reaction mixture for incorporating amino acids with S-30 extract and RNA phage messengers. The concentration of magnesium acetate necessary for optimal activity must be determined experimentally since it may vary with the particular strain used, the origin

1. The S-30 System from *Escherichia coli*

TABLE 2

Components of an Amino Acid-Incorporating Mixture Directed by Phage RNA Messenger

Component	Concentration
Tris-HCl (pH 7.8)	50 mM
NH_4Cl	60 mM
Magnesium acetate	7-10 mM
Reduced glutathione	10 mM
ATP-tris	1 mM
GTP	0.02 mM
Potasium phosphoenolpyruvate (PEP)	5 mM
Pyruvate kinase (PK)	30 µg/ml
[^{14}C]Valine, 25 Ci/mole	0.03 mM
19 Other [^{12}C]amino acids	0.05 mM
Phage RNA messenger	0.7 mg/ml
Incubated S-30 (iS-30)	0.2 vol
Unincubated S-100 supernatant	0.1 vol

and integrity of the messenger, the concentration of GTP, ATP, and PEP, and so on. However, since Capecchi [34] and Salser et al. [6] observed that fidelity of translation with natural messenger is maximal at a Mg^2 concentration lower than that required for maximal incorporation, it may be advisable to use a Mg^2 concentration slightly lower than the experimental optimum for incorporation. A concentration around 7.5-8.5 mM most frequently proves satisfactory.

The concentration of tris-HCl and NH_4Cl can be varied within reasonable limits without affecting the activity of the system. For instance, we did not observe a change in activity when the NH_4Cl concentration was lowered from 60 to 30 mM. However, increasing it above 100 mM was inhibitory. Na^+ ion is also inhibitory; thus it is preferable to use tris or K^+ salts of ATP and PEP rather than the Na^+ salts.

Reduced glutathione can be replaced by 6-10 mM 2-mercaptoethanol or 1-2 mM DTT as reducing agent. However, in the system described we have consistently observed that the last-mentioned two permit only 85-90% of the incorporation obtained with glutathione.

PEP and PK regenerate the ATP and GTP hydrolyzed by the amino acid-activating enzymes (ATP), by the binding of aminoacyl-tRNA and translocation (GTP), and by the very active ATPase and GTPase activities always present in crude extracts. The concentration of GTP (20 μM) is nearly 10 times lower than that used by most investigators. We have not observed a change in the activity of the system by increasing GTP concentration 10 times or by excluding it altogether. Presumably, as a result of the short dialysis time used in the preparation of the extract, enough endogenous GTP is present in the reaction mixture.

1. The S-30 System from <u>Escherichia coli</u>

[^{14}C]Valine is a convenient labeled amino acid for monitoring incorporation; it is one of the most common amino acids in the main product of the system (the phage coat protein [2-4,20]) and is fairly evenly distributed along the sequence of this protein [47]. However, other labeled amino acids suitable for incorporation studies can also be employed. Lysine, leucine, isoleucine, arginine, phenylalanine, and alanine are some of the labeled amino acids frequently used [2-4,14,22]. If incorporation into polypeptides other than the coat protein is to be measured, histidine should be used as this amino acid is the only one not present in the coat protein [20]. Amino acids present in large amounts in the coat protein should not be used at concentrations lower than the indicated (30 µM), since otherwise they might become limiting. The specific activity of the amino acid chosen should vary according to the size of the reaction mixture to be assayed, the isotope used, the efficiency of the counter, and so on. The specific activity indicated in Table 2 for [^{14}C]valine (25 Ci/mole) is given as a guideline. A 50-µl reaction mixture, which incorporates [^{14}C]valine to an extent similar to that indicated in Table 1, contains about 10^4 cpm of the amino acid in hot trichloroacetic acid (TCA)-precipitable form at a counting efficiency of 20%.

There is a linear relationship between incorporation and the amount of phage RNA added at low concentrations of this messenger. Higher concentrations are less effective in increasing the incorporation but, within practical limits, saturation or inhibition of the system by phage RNA is rarely observed. The curve should be determined for every preparation of phage RNA in order to obtain the most convenient compromise between incorporation and expenditure of phage RNA.

S-30 extract should be preincubated to deplete its content of endogenous messenger. This was done before the dialysis step in the original procedure of Nirenberg and Matthaei [1] during the preparation of S-30 extract. In our laboratory postponing preincubation until just before the use of the extract, as indicated by Capecchi [4] and Webster et al. [14], has always resulted in increased activity.

The usual ribosome concentration is approximately 1.6 mg/ml, about the highest that allows linearity between ribosome content and incorporation. Ribosome concentration in the final reaction mixture is adjusted either by adding small amounts of standard buffer to the iS-30 extract after preincubation (if only small adjustment is necessary), or by changing the relative proportions of iS-30 extract and S-100 supernatant (to keep the amount of soluble fraction constant when a large adjustment is required).

Addition of S-100 supernatant that has not been incubated stimulates the system between 30 and 100% and promotes the formation of polysomes. Addition of polyethylene glycol 6000,

1. The S-30 System from *Escherichia coli*

which stimulates the synthesis of T4 lysozyme [6], also stimulates the synthesis directed by RNA phage messenger. We do not include stripped tRNA in the reaction mixture since we have not observed a stimulation of the system by various tRNA preparations obtained from commercial sources. However, since many workers include its addition in similar systems, it may be advisable to check for possible stimulation by any tRNA preparation at hand.

Procedure

a. *Preparation of Stock Reagents.* Unless otherwise indicated the reagents are stable in solution for indefinite periods of time provided they are stored frozen. The following solutions should be prepared:

1 M Magnesium acetate

2 M Tris-HCl (adjusted to pH 7.8)

3 M NH_4Cl (no refrigeration required)

100 mM Potassium PEP (neutralize, pH paper)

100 mM ATP-tris (neutralize, pH paper)

10 mM GTP

1 M Reduced glutathione. Since the solution should be replaced about once a week, prepare a minimal amount of solution, e.g., 1 ml. To dissolve, neutralize with saturated tris-OH.

5 mM (each) 19 amino acids minus valine. Most easily prepared from 0.1 M stock solutions of the amino acids, mixing

1 vol of each of the 19 plus 1 vol of H_2O.

5% TCA (kept in refrigerator)

0.5% Bovine serum albumin

Prepare 1 ml of "energy", that is, a solution containing PEP, PK, ATP, and GTP at 10 times the concentration in the reaction mixture. This solution can be frozen and thawed many times without deterioration.

Prepare 1 ml of 20X concentrated salts solution, taking into account the contribution of magnesium acetate and NH_4Cl added to the reaction mixture by the iS-30 extract and S-100 supernatant. A 1-ml solution of concentrated salts can be prepared by mixing:

3 M NH_4Cl, 0.28 ml

2 M Tris-HCl, 0.50 ml

1 M Magnesium acetate, 0.09 ml (final magnesium acetate concentration in reaction mixture, 7.5 mM)

H_2O, 0.13 ml

b. <u>Preparation of Reaction Mixture</u>. The volume of the reaction mixture varies with the experiment to be performed. A 50-µl reaction mixture is prepared by mixing the following reagents in a Wasserman or similar tube kept at 0°C:

10 µl "Cocktail", that is, a mixture of salts, "energy," amino acids, and so on

15 µl H_2O or other additions whose influence on the reaction is to be determined

1. The S-30 System from Escherichia coli

10 μl mRNA

15 μl iS-30 plus S-100 supernatant

The "cocktail" is prepared by mixing:

5 μl 1 M Glutathione

5 μl 19 Amino acids minus valine (5 mM each)

5 μl 2 mM Valine (modify according to concentration and specific activity of [^{14}C]valine used)

10 μl 0.5 mM [^{14}C]Valine (75 Ci/mole; modify according to concentration and specific activity of [^{14}C]valine used)

25 μl Concentrated salt solution

50 μl "Energy"

The mRNA is not included in the "cocktail" since reactions without messenger should always be run as controls and this incorporation is subtracted. It is also convenient not to add the viral messenger until just before the addition of the bacterial extract because of its extreme lability toward contaminating traces of RNase. (Caution: human fingers and sweat contain high RNase activity.)

S-30 extract is incubated, just before its addition to the reaction mixture, for 7 min at 34°C after it is mixed with 0.02 vol of 2 M tris-HCl (pH 7.8) and 0.1 vol of "energy." After the preincubation step, the iS-30 extract is chilled and mixed with 0.5 vol of S-100 supernatant and, if necessary, with a small amount of standard buffer to adjust the final ribosomal concentration.

c. _Incubation_. Since initiation of polypeptide synthesis with natural messenger is very sluggish at 0°C (Section VII,D), no polymerization takes place while the reaction mixture is kept at this temperature. Thus the reaction is normally started by placing the tubes into a 34°C bath. (When early kinetics of incorporation are studied, the reaction should be started by the addition of messenger to a preheated reaction mixture.) After a delay of less than 1 min linear incorporation begins and continues for approximately 10 min, when the incorporation reaches between 70 and 80% of its maximal value. Then incorporation usually slows down and stops at any time between 15 and 30 min, depending on the particular extract used and other undefined variables. It follows that if reaction rate, rather than total incorporation, is to be measured, the reaction should be stopped before 10 min. When a high degree of precision is not required, advantage can be taken of the fact that in most cases the reaction rate is roughly proportional to the total incorporation. Incubating tubes at a temperature lower than 34°C slows down the rate of synthesis and prolongs the linear phase of incorporation [23]. However, total incorporation is usually less than at 34°C. After incorporation is completed incubation should not be prolonged since proteolytic enzymes can attack the synthesized product rendering it soluble in hot TCA.

1. The S-30 System from _Escherichia_ _coli_

d. Determination of Incorporated Radioactivity. The reaction is usually terminated by chilling the tube in an ice-water bath and precipitating the proteins with the addition of a 10-fold excess of 5% TCA. A 50-µl reaction mixture contains sufficient protein and RNA to produce a good-sized precipitate. However, when a smaller portion of the reaction mixture is analyzed (20, 15, and even 10 µl are sufficient for the determination without unduly affecting the precision of the results), it is convenient to mix it with 1 drop of 0.5% bovine serum albumin before TCA precipitation. The mixture is then heated to 90°C for 20 min to hydrolyze precipitated aminoacyl-tRNA, chilled again, and filtered with suction through a suitable filter (Millipore HA, Gelman GA, Sartorius, or Oxoid membrane filters, Whatman glass fiber, and so on). The tube is carefully rinsed two or three times with 2- or 3-ml portions of 5% TCA to remove any adherent particles of precipitate. The walls of the filtration apparatus in contact with the mixture are also rinsed once with a stream of 5% TCA from a wash bottle. Depending on the particular isotope incorporated, the filters can either be dried and counted by liquid scintillation, or glued while wet onto planchets, dried, and counted in a Geiger or gas-flow counter.

Remarks. Under optimal conditions we have observed incorporations as high as 30 molecules of valine per ribosome, equivalent to the synthesis of approximately 2 molecules of phage coat protein per ribosome. Somewhat higher incorporations (about 3 molecules of phage coat protein per ribosome) have been obtained by others with very similar systems (M. Capecchi, personal communication, and Ref. 14). These values represent well over 100-fold stimulation above background incorporation by the addition of phage messenger (Table 1). It is clear that these S-30 systems are active in the three stages of protein synthesis: initiation, elongation, and termination. Furthermore, ribosomes that have completed synthesis of a given protein can start at least a second round of protein synthesis.

Preincubation of S-30 extracts for periods longer than 7 min, as recommended in many procedures, does not substantially decrease the residual endogenous activity (Table 1) and results, in our laboratory, in decreased activity of the extract.

When isolated ribosomes are used for incorporation studies with crude systems, it is a good practice to mix them with S-100 supernatant. After mixing one can proceed as with an iS-30 plus S-100 mixture. In general, activity with these reconstituted mixtures is lower than with iS-30 extracts. However, there have been reports of reconstituted systems with purified ribosomes that are as active with natural messenger as crude iS-30 extracts [48].

If synthesis is directed with T4 messenger, rather than with RNA phage messenger, the same reaction mixture can be

1. The S-30 System from *Escherichia coli*

used [6]. With this messenger one can assay for incorporated radioactivity and for synthesized active lysozyme. Lysozyme activity is measured by either the decrease in absorbance of a suspension of chloroform-treated *E. coli* cells [6,49] or, more easily, by the loss of radioactivity from *E. coli* cell walls grown with labeled 1,6-diaminopimelic acid (a precursor of cell-wall mucopeptide) and stuck to discs of Whatman 3MM filter paper [50].

The reaction mixture directed by endogenous messenger is set up as for exogenous natural messenger, but preincubation of S-30 extract is omitted. DNase treatment of S-30 extract should be avoided if de novo synthesis of messenger in the system is to take place. In any case, incorporation is usually complete in 10-15 min [21]. Incorporated radioactivity is determined as above.

The most common cause of problems in incorporation experiments is defective S-30 extract, S-100 supernatant, or messenger. Sometimes for no apparent reason a particular S-30 extract or S-100 supernatant becomes turbid, because of the appearance of a precipitate, either shortly after it is thawed, or during or after preincubation. This extract shows a much decreased activity and should not be used. Sometimes precipitation of an S-30 extract after preincubation can be avoided by rapidly mixing it with S-100 supernatant. Decreasing the time of preincubation may also help. Still, the replacement of the defective extract is the solution of choice.

Partially degraded messenger is another common cause of low activity. Exhaustive precautions should be taken during its preparation (sterile glassware, the use of gloves, and so on). Not all methods used for the purification of the phage and extraction of the RNA yield preparations of equal activity. Liquid-polymer partition methods for concentration of the phage [51], and CsCl equilibrium centrifugation for its purification [52] are, in our experience, quite satisfactory. RNase inhibitors such as bentonite or Macaloid (National Lead Company, Houston, Texas) should be used during cold phenol extraction of RNA [52].

B. Synthesis with Synthetic Polynucleotides

The most commonly used synthetic polynucleotides are the homopolymers: poly U, poly C, poly A, and poly I. They direct the synthesis of polyphenylalanine, polyproline, polylysine, and a copolymer of valine plus glycine, respectively [53]. Synthesis is carried out in reaction mixtures set up as for natural messenger but with some modifications. The amino acids not coded for by the homopolynucleotide can be omitted. The labeled amino acid, at a specific activity between 10 and 20 Ci/mole, should not be at a concentration lower than 0.02 mM. The Mg^2 concentration for optimum activity, which must be determined experimentally, usually lies between 12 and 18 mM

1. The S-30 System from Escherichia coli

for poly U, poly C, and poly A, and between 17 and 30 mM for poly I [36]. NH_4Cl concentrations higher than 60 mM have been used for maximal activity [29,33]. In our experience, supplementing S-30 extract with S-100 supernatant enhances the activity with poly U, but to a lesser extent than with natural messenger. In contrast to the natural system, addition of tRNA is normally stimulatory with the homopolynucleotides [53]. With poly U, poly C, and poly I, tRNA is used at concentrations from 0.2 to 0.5 mg/ml; with poly A concentrations of about 1 or 2 mg/ml are recommended [53]. The appropriate concentration of the homopolynucleotide, usually between 0.05 and 0.4 mg/ml, should be experimentally determined. When poly I is used, it should be added to the reaction mixture last, since otherwise it precipitates because of the high Mg^2 concentration and becomes inactive [36]. Since synthesis continues longer than with natural messenger, incubation should be prolonged to 30 or 60 min (or even 90 min with poly I). The reaction can be stopped with 5% TCA except when poly A is used. Polylysine peptides are soluble in 5% TCA [54]. They are precipitated by a fresh mixture of 0.25% sodium tungstate and 5% TCA (final pH 2.0) [55]. Some workers include carrier polylysine during precipitation [55], while others do not [53,56]. Since small lysine peptides do not precipitate easily [46], the addition of carrier may enhance their recovery. Filters should be washed with TCA-tungstate.

Details on synthesis directed by other synthetic polynucleotides can be found in appropriate references [22,31,32, 36,55,57-60].

VII. ANALYSIS OF POLYPEPTIDE-SYNTHESIZING SYSTEMS IN SUCROSE DENSITY GRADIENTS

Sucrose density gradient analysis of an amino acid incorporation mixture is a powerful tool that allows us to probe into many aspects of the synthesis. For instance, the ribosomes are separated from the remaining low-molecular-weight components of the system and further fractionated into subunits, free 70 S ribosomes, 70 S monosomes (combined with messenger), and polysomes. The synthesized product can be separated into released, presumably completed, polypeptide chains and nascent incomplete chains still attached to ribosomes. Each class of ribosomal subunit and ribosomal aggregate, as well as its associated nascent chains, can be quantitated. The number of ribosomes active at any time during the synthesis can be estimated. The main product of the system directed by RNA phage messenger (the coat protein of the phage) can be easily visualized in gradients by its specific attachment to the free messenger. Effects of antibiotics on polypeptide chain initiation, chain extension, and polysomal stability can be very easily determined.

1. The S-30 System from Escherichia coli

   A. Preparation of Sucrose Density Gradients

   A 15-30% linear sucrose density gradient in standard buffer (usually minus 2-mecaptoethanol) is appropriate for the analysis of the reaction mixture. In our laboratory this gradient allows better resolution than the commonly employed 5-20% gradient. 3.8 ml (for an International SB-405 rotor) or 4.8 ml (for Spinco SW39, SW50L, and SW50.1 rotors) gradients are the most convenient to use for analytical purposes. For preparative purposes larger gradients should be used.

   Linear gradients are easily prepared (see also Chapter 5) with the help of a commercially available apparatus consisting of two communicating vessels of transparent Lucite drilled in a block as shown in Fig. 1. A volume of dense sucrose solution (30%) in standard buffer equal to half the gradient size is placed in well A, while the screw clamp D is tight and the stopcock C is closed. An equal volume of light sucrose solution

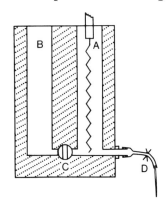

Fig. 1. Sucrose density gradient-forming apparatus. See text for description.

(15%) in standard buffer is placed in well B (these solutions should be stored in a refrigerator). Well A is provided with a motor-driven stirring device (a magnetic stirrer fitted with a small magnet or piece of wire is also suitable). The tip of the exit needle is placed slightly below the border of the centrifuge tube touching the inner wall so the sucrose solution can slide smoothly to the bottom of the tube. The stopcock C is opened and any air lodged in the tube between the two wells is displaced by slight pressure on well B (by covering it with a finger). The screw clamp D is then loosened to let the sucrose solution run down to the tube. It should take several minutes for the sucrose solution to run into the tube. Larger gradients should take a proportionally longer time to form. In our experience it seems to make no difference whether gradients are formed in a cold room with chilled sucrose solutions or are chilled after they have been formed at room temperature. If necessary, linear gradients formed in this way can be used immediately after chilling (the common practice is to let them stabilize for a few hours in the cold).

When many gradients must be prepared, it is more convenient to form them by layering four equal volumes of 30, 25, 20, and 15% sucrose solutions in centrifuge tubes. These gradients should be left to equilibrate overnight in the cold. They provide analyses indistinguishable from those obtained with gradients formed with the apparatus described above.

1. The S-30 System from *Escherichia coli*

B. Zonal Centrifugation

A 30- or 40-μl portion of the reaction mixture to be analyzed, equivalent to about 1 A_{260} unit of ribosomes, is layered on a gradient. To stop further reaction the gradient can be layered with 40 μl of 0.2% chloramphenicol or another suitable inhibitor. We have observed [23] that chloramphenicol does not induce polysome formation in unincubated reaction mixtures but slightly improves the recovery of polysomes and nascent chains attached to the ribosomes in incubated mixtures, presumably because it prevents run-off. As a guideline, centrifugation at 5°C for 110 min at 39,000 rpm in a Spinco SW39 rotor, or 75 min at 45,000 rpm in a Spinco SW50.1 rotor, or 60 min at 60,000 rpm in an International SB-405 rotor brings the 70 S ribosomes toward the center of the gradient and the polysomal tetramers close to the bottom. Since the velocity of sedimentation is very much affected by the temperature, which is difficult to control with precision in most ultracentrifuges, the most convenient time is dictated by experience. Once the correct time is known for a given speed, temperature, and rotor, the centrifugation time for a different speed (if the other conditions are the same) can be calculated by the relationship:

$$t_1 \, (\text{rpm}_1)^2 = t_2 \, (\text{rpm}_2)^2.$$

It is a common practice to decelerate an ultracentrifuge without a brake. However, a brake may be used if very high resolution is not essential, and provided it functions properly. A faulty brake can completely ruin a set of gradients, while it is completely satisfactory in conventional centrifugation. Gradients should be kept at 0°C after centrifugation and analyzed as soon as possible. Delays of as little as 1 or 2 hr cause visible broadening of the peaks.

C. Analysis of Gradients

Sucrose density gradients are traditionally analyzed by piercing the bottom of the tubes with a short hypodermic needle, fitted with a suitable sealing assembly, and slowly emptying the tubes through it. The flow rate can be controlled by a syringe connected to the top of the gradient through a perforated stopper and a short piece of flexible tubing. Fractions of several drops are collected and their absorbance, after suitable dilution, and radioactivity determined.

Except for some special purposes, this method is currently being replaced by the use of an automatic gradient analyzer (ISCO, Gilford). This apparatus consists of a device for displacing the gradient from the tube and making it pass through the flow cell of a spectrophotometer whose output is connected to a recorder. A continuous scan of the absorbance of the gradient is obtained. In addition to being a saving in labor, a gradient analyzer permits a much greater resolution in the analysis. This is attributable in part to the special design of the components through which the gradient moves (designed

1. The S-30 System from Escherichia coli

to minimize turbulence), and in part to its ability to measure absorbance continuously in the undiluted gradient. This permits the use of smaller amounts of material for the analysis and prevents overloading of the gradient.

Figures 2 and 3 show A_{254} profiles of gradients that contain reaction mixtures directed with R17 RNA. They were obtained with an ISCO Model 180 gradient fractionator and Model 222 UV analyzer. This apparatus displaces the gradient upward by injecting 50% sucrose into the bottom of the tube. The gradient moves through a short, tapered canal, which leads directly into the flow cell. The cell with the shortest length of light path (2 mm) is routinely used since it permits higher resolution. The flow rate is adjusted so that it takes about 10 min to scan a 5-ml gradient. Higher flow rates should not be used with small analytical gradients, not even to raise the gradient up to the flow cell (this results in exaggerated trailing of the material from the top of the gradient, which unduly raises the base line of the immediate ribosomal peaks). After leaving the flow cell, the gradient can be fractionated in, for example, 5-drop fractions (Figures 2 and 3), which can be analyzed for radioactivity or any other desired parameter. If total radioactivity in each fraction is to be determined (Figure 2), drops can be collected directly into scintillation vials. Five drops, plus 0.9 ml of H_2O, plus 10 ml of Bray's scintillation fluid is appropriate (water must be added to keep

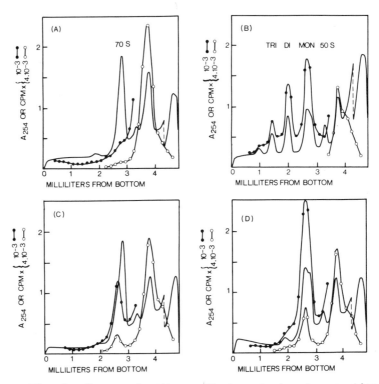

Fig. 2. Sucrose density gradient analysis of a reaction mixture as in Table 2 but with [^{14}C]valine substituted by R17[^{32}P] RNA. Incubations at 34°C. Forty-μl portions were centrifuged in 4.8-ml gradients in a SW50.1 Spinco rotor for 75 min at 45,000 rpm. Gradients were analyzed in an ISCO gradient analyzer and 5-drop fractions were collected and analyzed for total radioactivity. (A) Complete reaction mixture, unincubated. (B) The same, incubated 6 min. (C) The same, incubated 5.5 min in the presence of 20 μg/ml of streptomycin. (D) The same, incubated 5.5 min in the presence of 100 μg/ml of chloramphenicol.

1. The S-30 System from Escherichia coli

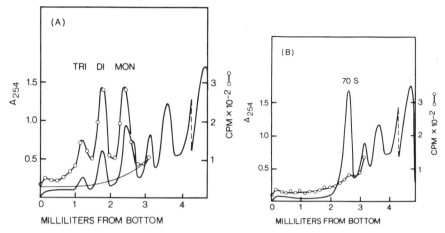

Fig. 3. Sucrose density gradient analysis of a reaction mixture as in Table 2. Specific activity of the [^{14}C]valine was 66 Ci/mole. Incubation, centrifugation, and analysis as in legend to Figure 2, except that hot TCA-precipitable radioactivity was determined. (A) Complete reaction mixture incubated for 7 min. [Thin line across polysomes corresponds to radioactivity level observed in (B).] (B) The same, but further incubated for 3 min in the presence of 100 µg/ml of puromycin.

the sucrose in solution). Addition of a thixotropic gelling agent (Cab-o-sil) may be helpful when larger fractions are collected. If cold or hot TCA-precipitable radioactivity is to be determined, 1 drop of 0.5% bovine serum albumin, or other appropriate carrier, and 1 ml of 5% TCA are added to each fraction. The mixtures can be filtered after a few minutes (cold TCA-precipitable radioactivity) or heated at 90°C for 20 min before filtration (hot TCA precipitable radioactivity) (Figure 3). To determine hot TCA precipitable radioactivity, the tubes are shaken immediately before they are placed in the hot-water bath to resuspend the more-or-less settled precipitate. Otherwise, because of the presence of sucrose the precipitate clumps together in a few large pieces upon heating, and the counting efficiency is altered as a result of self-absorption.

D. Interpretation of Gradient Profiles

Figure 2 shows the A_{254} and total radioactivity profiles of an unincubated complete amino acid-incorporating system such as that described in Table 2 but with the labeled [^{14}C]valine replaced by R17[^{32}P]RNA. Sedimentation is from right to left. The A_{254} profile shows a large peak at the top of the gradient (the sensitivity of the recorder was cut in half to avoid having the peak run off the chart) as a result of the relatively low-molecular-weight UV-absorbing materials in the reaction mixture (proteins, tRNA, ATP, GTP, and so on). The next large peak contains a small peak of 30 S ribosomal subunits completely masked by the free R17 RNA, which sediments at 27 S. The third large peak corresponds to the bulk of the ribosomes, which sediment at 70 S. In between the 70 S ribosomes and the messenger peaks, there is a small peak corresponding to the 50 S subunits. The radioactivity profile shows that

1. The S-30 System from *Escherichia coli*

most of the messenger is undegraded and in the free form. In this unincubated mixture, almost no 70 S ribosomes are combined with messenger, indicating that initiation is very sluggish at 0°C. The absorbance profile also shows, two-thirds of the way down the gradient, a small peak that corresponds to aggregates of two 70 S ribosomes sedimenting at 100 S [61]. Since there is no radioactivity associated with them, they may represent pairs of 70 S ribosomes held together by ionic interactions and/or residual dimers resulting from the fragmentation of polysomes during alumina grinding and therefore held together by endogenous mRNA.

Analysis of the same reaction mixture incubated for 6 min at 34°C is shown in Figure 2B. The absorbance profile shows that a large fraction of the ribosomes now sediment as aggregates larger than 70 S, mostly as dimers, trimers, and tetramers [4,34a]. They are combined with R17[^{32}P] RNA, and it is shown below that they are active polysomes. We have not been able to detect with this technique polysomes larger than pentasomes. This may reflect an intrinsic limitation of the messenger, since it has been reported that in cells infected with MS2 phage most of the single-stranded RNA is associated with polysomes that contain two to five ribosomes [62]. The monomer peak at 70 S is now smaller and disproportionately broader. It is a composite of two peaks: free 70 S ribosomes and 70 S ribosomes complexed with R17[^{32}P]RNA (monosomes), which sediment slightly faster than 70 S [34a]. The free 70 S peak can be seen as a shoulder on the larger peak of complexed 70 S (see also Figure 3A, and Figure 3 in Ref. 63). As expected, there is no radioactivity associated with the 50 S peak. The composite peak at 30 S is smaller because of the depletion of free R17 RNA; part of it is combined with the ribosomes and another part, as indicated by the radioactivity profile, has been degraded by RNase. Approximately 36% of the total radioactivity sediments beyond the valley between the 50 and 70 S peaks. It can be taken as an indication of the amount of messenger bound to the ribosomes.

Figure 2C shows an analysis of the same reaction mixture incubated at 34°C for 5.5 min but in the presence of streptomycin, an antibiotic that directly inhibits polypeptide chain elongation and also inhibits polypeptide chain initiation by promoting breakdown of the 70 S initiation complex [23,64]. The absorbance and radioactivity profiles are almost identical to those of the unincubated control (Figure 2A) except for the presence of a small amount of ribosomes combined with R17[^{32}P]RNA, which appear as a shoulder on the leading edge of the large free 70 S peak. Polysomes are absent. This figure can be compared with Figure 2D which shows the analysis of an identical mixture, but with streptomycin replaced by chloramphenicol, an inhibitor of peptide bond formation [40]. Polysomes are almost absent, but a large proportion of the 70 S

1. The S-30 System from Escherichia coli

ribosomes are now complexed with R17[^{32}P]RNA. In fact, 28% of the total radioactivity sediments as a complex with ribosomes. This indicates that chloramphenicol does not interfere with polypeptide chain initiation, in agreement with the known mode of action of the antibiotic.

When the label in the reaction mixture is not in the messenger but in an amino acid (valine), a gradient analysis similar to that in Figure 3A can be obtained. The complete reaction mixture was incubated for 7 min at 34°C in the absence of antibiotics before it was placed on a gradient already layered with chloramphenicol to stop any further reaction (Section VII,B). There is hot TCA-precipitable radioactivity associated with monosomes and polysomes. It is not shown, but radioactivity is also found sedimenting near the top of the gradient and associated with the 30 S peak [4,23]. It represents released, completed protein molecules; most of them are coat protein of the phage, which sediments at 30 S because of the formation of a complex with the viral messenger [4]. In the experiment shown in Figure 3A, released radioactivity amounted to 50% of the total hot TCA-precipitable radioactivity in the gradient.

1. Estimation of Number of Active Ribosomes

The ribosomal aggregates seen in the absorbance profile of Figure 3A are mostly active polysomes since (1) they possess mRNA in approximately the expected ribosome-to-messenger ratios (unpublished observations); (2) the average length of their radioactive nascent chains is that expected of a system synthesizing protein in steady-state conditions (see below); and (3) puromycin rapidly and completely converts them to free 70 S ribosomes (Figure 3B and Refs. 23 and 63). (This antibiotic is an analog of aminoacyl-tRNA; it accepts the transferred peptidyl chain, causing its premature release [40]. As a consequence, the ribosome falls off the messenger.) Therefore, determination of the number of ribosomes in polysomes, plus the monomers combined with messenger, provides an estimate of the number of ribosomes active in polypeptide synthesis.

The distribution of ribosomes is estimated by cutting out and weighing the different peaks in the graph (or a suitable copy). Since the free viral messenger sediments with the 30 S subunits, the 30 S peak can be assumed to be one-half the 50 S peak. To correct for nonribosomal material, the base line for the polysome analysis is determined from an analysis, run in parallel, of a sample of the reaction mixture incubated for a few minutes with puromycin (Figure 3B).

The fraction of active ribosomes in the monomer peak (complexed monomers) can be estimated from the absorbance profile, provided that the free and complexed 70 S peaks are sufficiently resolved. For instance, the monomer peak in Figure 3A was estimated to contain 60% of complexed (active)

1. The S-30 System from Escherichia coli

ribosomes. When resolution is poor, the fraction can be estimated from the radioactivity associated with this peak (see below), by the assumption that active monomers and dimers possess the same specific activity. This procedure may slightly underestimate the number of active monomers, since they possess a somewhat lower specific activity than the dimers (Section VII, D,3 and Ref. 34a). If desired, the weight of each polysomal peak can be corrected for the absorbance of the messenger associated with the ribosomes. This is easily done by assuming that (1) each active monomer or polysome contains an intact molecule of messenger, and that (2) the ribosomal RNA and phage RNA possess the same specific absorbance at 254 nm. It follows that the correction factors are equal to the molar fraction of ribosomal RNA in each ribosomal-messenger aggregate. Since the molecular weight of, for instance, R17 RNA is 1.1×10^6 [52] and a 70 S ribosome contains about 1.8×10^6 daltons of RNA [61], the molar fractions are: 0.62 for monosomes, 0.77 for disomes, 0.83 for trisomes, and 0.87 for tetrasomes. When this procedure was applied to the reaction mixture of Figure 3A, it was estimated that 44% of the ribosomes were engaged in synthesis at the moment the sample was taken. They were distributed as follows: 46% in monosomes, 40% in disomes, and 14% in trisomes. As many as 60% of the ribosomes in very active systems may be engaged in synthesis at one time.

2. Estimation of Amino Acid Incorporation Rate

The amino acid incorporation rate can be calculated if the percent of active ribosomes and the rate of incorporation of [^{14}C]valine is known. If we assume that the product of the system (mainly coat protein) contains 10% valine [20], the following relationship can be established:

Average amino acid incorporation rate (seconds) =

$$\frac{\text{Percent active ribosomes X [ribosomes] X S.A. X [}^{14}\text{C]valine}}{\text{Incorporation rate X 2.7 X 10}^9}$$

The final concentration of ribosomes in the reaction mixture should be expressed in micrograms per milliliter; the specific activity of [^{14}C]valine in counts per minute per μmole; and the incorporation rate in counts per minute of [^{14}C]valine incorporated per second per milliliter of reaction mixture. Applying the experimental values for the reaction mixture of Figure 3A, we find that average amino acid incorporation rate at 34°C is:

$$\frac{44 \times 1440 \times 29.4 \times 10^6}{593 \times 2.7 \times 10^9} =$$

1.2 sec per amino acid per ribosome.

This value closely agrees with estimates made by another procedure [65]. At 14°C the time increases to approximately 13 sec per amino acid per ribosome [23]. Amino acid incorpor-

1. The S-30 System from Escherichia coli

ation rate measurements have helped to demonstrate conclusively that the antibiotic streptomycin inhibits polypeptide chain elongation when added to an actively synthesizing system [23]. The method can be applied to any other antibiotic.

3. Estimation of Average Polypeptide Chain Length

The radioactivity associated with each polysomal peak (Figure 3A) should be corrected for the residual base line radioactivity observed in the monosome-polysome region in the gradient from a reaction mixture treated with puromycin (Figure 3B) (see thin line across polysomes in Figure 3A). This residue probably represents incorporation associated with particulate materials (e.g., membrane fragments) present in the S-30 extract. The corrected values, expressed in picomoles of [^{14}C]valine divided by picomoles of ribosomes in each peak, yield the average polypeptide chain length expressed in molecules of [^{14}C]valine per ribosome. The values for the polysomal peaks of Figure 3A are 5.0 for monosomes, 6.0 for disomes, and 10.7 for trisomes (figures corrected for the incomplete recovery, about 80% of hot TCA-precipitable radioactivity from the whole gradient). It seems clear that the average length of the nascent polypeptide chain increases with the size of the polysome, as previously observed by Engelhardt et al. [34a]. Considering that the main product of the system, the coat protein of the phage, contains 14 valine residues per polypeptide

[20], these values support the presumption of steady-state synthesis of this protein during the middle portion of the linear incorporation phase. They also support the contention that most ribosomes in the polysomes are active.

VIII. EXTRACTION OF POLYSOMES FROM E. COLI CELLS

In E. coli cells protein synthesis takes place on large aggregates (polysomes) consisting of a molecule of mRNA with many attached ribosomes. In rapidly growing cells the majority of the ribosomes are actively synthesizing protein in polysomes. The remaining ribosomes, as observed in extracts, are partly in the form of native or immature subunits, and partly as 70 S monomers. These monomers may be either polysomal monomers produced by the breakage of polysomes, 30 S-50 S couples present in the cell and devoid of mRNA and peptidyl-tRNA (that is, free 70 S ribosomes), or couples of 30 and 50 S subunits artifactually formed during cell lysis. The exact nature of 70 S ribosomes in extracts, that is, whether they exist in vivo as 70 S monomers or as 30 and 50 S subunits and polysomes only, is controversial [66-83].

The methods used to disrupt E. coli cells that are mild enough to preserve (as well as possible) the in vivo distribution of ribosomes are based on a two-step strategy. The cell wall is attacked either by an EDTA-lysozyme or freeze-thaw-lysozyme treatment, or by growing the cells in a fragile form (growth

1. The S-30 System from Escherichia coli

in the presence of a high concentration of salts or use of appropriate fragile mutants). In a second step cells are lysed by treatment with a detergent, usually deoxycholate (DOC) or the nonionic compounds Brij 35 or 58 (Atlas Chemical Industries, Wilmington, Delaware). The extracts thus obtained, after treatment with DNase, are analyzed in a sucrose density gradient or used for in vitro protein synthesis with endogenous messenger. The methodology is well documented in the literature [66,72,76, 84-87]. Here the freeze-thaw-lysozyme and EDTA-lysozyme procedures, as used by Ron et al. [84] and Godson and Sinsheimer [85,86], respectively, are detailed. Other published procedures are briefly discussed.

A. Cell Disruption by Freeze-Thaw-Lysozyme Treatment

This method capitalizes on the fact that freezing and thawing make E. coli cell walls susceptible to attack by lysozyme. It is usually applied to small cultures (30-100 ml), but there is no apparent reason why it could not be scaled up to disrupt the cells from larger cultures. Escherichia coli cells growing at mid-log phase (2-4 X 10^8 cells per milliliter) in 100 ml of minimal medium plus glucose [88] or broth are chilled, as quickly as possible, by pouring the culture onto an equal amount of ice that has been cooled at -20°C, and the mixture is centrifuged immediately at 10,000 X g for 5 min. The supernatant is discarded, residual medium in the tube is aspirated with a Pasteur pipet, and the cells are quickly resuspended in 1 ml of cold buffer (60 mM NH_4Cl or KCl, 10 mM magnesium acetate, 10 mM tris-HCl (pH 7.8); see Section IV) to which 50 µl of a 20 mg/ml solution of lysozyme (which should be stored in a freezer) is added. The suspension, in a narrow plastic centrifuge tube, is quickly frozen in an acetone-dry ice mixture and then slowly thawed in a beaker that contains fairly cool water (about 5-10°C) and is provided with a magnetic stirrer capable of whirling the tube around until the last bit of ice is melted. After a second cycle of freezing and thawing, cells are lysed at 0°C by the addition of 20 µl of a neutral solution of 10% sodium DOC (final DOC concentration, 0.2%). In a few minutes lysis is complete. Further cycles of freezing and thawing beyond the second do not improve the yield of RNA significantly [84].

The lysate is very viscous at this stage because of the presence of DNA. DNA and cell debris can be eliminated by centrifugation at 30,000 X g for 10 min. However, since some polysomes, especially large ones, are trapped in the pellet, it is best to first digest the DNA by the addition of DNase (20 µl of a 1 mg/ml solution). After a few minutes at 0°C, the extract, without clarification, is ready for analysis in a sucrose density gradient. It is a good practice to handle the lysate gently and with wide-bore pipets to avoid breakage of polysomes by shearing. Cell debris is found at the bottom of

1. The S-30 System from *Escherichia* coli

the gradient as a thin pellet and does not interfere with the analysis.

Sucrose density gradient analysis is performed as previously described (Section VII). Linear 15-30% sucrose gradients in standard buffer are appropriate. The portion of the extract to be analyzed depends on the number of cells extracted, the size of the gradient, the sensitivity of the gradient analyzer, and so on. With an ISCO gradient analyzer fitted with a 2-mm cell (light-path length), 50-100 μl of extract are normally sufficient for a 4.8-ml gradient; 26-ml gradients (Spinco SW25 rotor) require proportionally larger amounts. Approximate centrifugation times and velocities required to sediment the largest polysomes close to the bottom of the tube are: 30 min at 45,000 rpm in an SW50.1 rotor; 45 min at 39,000 rpm in an SW39 or SW50L rotor; and 2.5 hr at 25,000 rpm in an SW25 rotor. The A_{260} profiles obtained from fast-growing cells usually show a large peak on top of the gradient, immediately followed by the 30, 50, 70 S, and polysomal peaks (Figure 4). The 70 S peak may be of approximately the same size, smaller, or larger than the 50 S peak depending on the strain and medium used, the amount of polysomal run-off that may have taken place during preparation of the extracts, and so on. To prevent run-off, chloramphenicol (100 μg/ml) can be added to the culture immediately before chilling. Additional antibiotic should be present in the resuspension

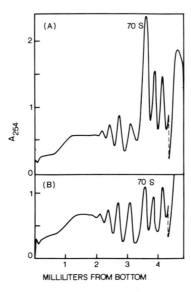

Fig. 4. Sucrose density gradient analysis of a polysomal extract from E. coli B. Cells in 50 ml of minimal medium with glucose [88] were grown to a density of 2×10^8 cells per milliliter. They were disrupted by the freeze-thaw-lysozyme technique. 200 µl of lysate were centrifuged in 4.7-ml gradients for 30 min at 45,000 rpm in a SW50.1 Spinco rotor. Analysis was in an ISCO gradient analyzer. Larger polysomes reached the bottom of the tube. (A) Lysate prepared without chloramphenicol. (B) Lysate prepared with 80 µg/ml of chloramphenicol.

buffer. Comparing Figures 4A and 4B, we see that chloramphenicol sharply decreased the size of the 70 S peak (and, to a lesser extent, the subunit peaks) and preserved the larger polysomes. From the absorbance profile it is easy to estimate the number

1. The S-30 System from Escherichia coli

of ribosomes in polysomes (Section VII,D,1). The polysomal base line can be precisely determined from a parallel gradient with a portion of the extract incubated for several minutes at 0°C with 5 µg/ml of pancreatic RNase; this treatment should convert all polysomes to 70 S monomers [84].

Remarks. This method has been applied to E. coli B, W, and some K12 strains with consistently successful results. With other strains success seems to be very dependent on the exact conditions of freezing and thawing. Slow thawing, by the use of narrow plastic tubes and quite-cold water, appears to be required to extract the larger polysomes. This method may not be easily adapted to some other strains, as is also true of other methods [86]. With a particular strain, success may depend on the growth medium used and the age of the culture. Early-log phase cells appear to be the easiest to disrupt. According to Ron et al. [84], essentially all of the cells are disrupted and more than 75% of the total RNA is extracted by this technique.

Since polysomes are easily dissociated by run-off in the whole cell when growth is interrupted (especially in the absence of chloramphenicol) but are relatively stable in the cold lysate, fast cooling of the culture and disruption of the cells as quickly as is compatible with the procedure is recommended. This precaution applies to most techniques.

Caution should also be exerted when varying the Mg^2 concentration of the lysate fluid. Its influence on the extraction of polysomes by the freeze-thaw-lysozyme technique has not been documented, but Mg^2 concentration is known to be critical in the EDTA-lysozyme method (see below and Refs. 85 and 86).

The main advantages of the freeze-thaw-lysozyme method are: (1) its avoidance of EDTA, which facilitates the degradation of mRNA by RNases in the cell by its chelation of Mg^2 (low concentrations of Mg^2 activate RNases); (2) the possibility of processing many samples simultaneously (up to six tubes can be thawed at the same time in a 250-ml beaker containing about 100 ml of cold water); and (3) the possibility of storing the cells frozen at -70°C before the addition of DOC to perform sucrose gradient analysis at the investigator's convenience. Storage for up to several days does not result in degradation of polysomes [84]. The disadvantages of the freeze-thaw-lysozyme method, when compared to the EDTA-lysozyme technique, are: (1) its relative slowness, which may promote polysome run-off, and (2) the high concentration of DOC necessary to lyse the cells. Since DOC inhibits protein synthesis (our unpublished observations) the extracts may be unsuitable for amino acid incorporation or in vitro polysome metabolism studies.

B. Cell Disruption by EDTA-Lysozyme Treatment

This method is based on the fact that EDTA treatment renders E. coli cell walls susceptible to attack by lysozyme.

1. The S-30 System from Escherichia coli

Typically, cells from a 100-ml culture are grown, chilled, and collected by centrifugation as described for the freeze-thaw-lysozyme technique. Godson and Sinsheimer [85] recommend cooling the culture by pouring it into a 600-ml beaker immersed in an acetone-dry ice bath. The culture is swirled to prevent icing. In less than 10 sec, the culture should be at 2-4°C; it is immediately transferred to centrifuge tubes before any freezing occurs. All the remaining operations are carried out at 0°C. The cells are resuspended in 0.4 ml of 25% sucrose containing 10 mM tris-HCl (pH 8.1); 0.1 ml of an ice-cold mixture of equal parts of 0.85 mg/ml crystalline lysozyme in 0.25 M tris-HCl (pH 8.1) and 8 mM EDTA is added. After 45 sec to 2 min (see Remarks), the cells are mixed with 0.5 ml of an ice-cold lytic mixture consisting of 0.1 ml of 5% (w/v) Brij 35 [the original method [85,86] specified Brij 58 (polyoxyethylene cetyl ether), which can precipitate at 0°C; Brij 35 (the lauroyl ether) is preferred [68,83] (see Chapter 5)] in 10 mM tris-HCl (pH 7.2), 0.1 ml of 100 mM magnesium acetate, 0.2 ml of 1% DOC (see Remarks) in 0.1 M tris-HCl (pH 8.1), 0.025 ml of 1 mg/ml DNase, and 0.025 ml of H_2O. (Both mixtures should be prepared just before starting the experiments. Stock solutions of the components can be stored frozen.) After 1-2 min most of the turbidity of the suspension should disappear and the cells have lysed. In addition to other components, the lysate contains: 10 mM acetate, 0.5% Brij 35, 0.2% DOC, 10% sucrose,

40 µg/ml lysozyme, 0.4 mM EDTA, and 25 µg/ml of DNase. The lysate can be analyzed in sucrose density gradients as described (Section VIII,A). Polysome patterns from fast-growing cells such as those shown in Figure 4 are routinely obtained.

Remarks. According to Godson and Sinsheimer [85], the method can be applied to practically any E. coli strain. However, since fragility of the cells varies with strains and conditions of growth, the procedure must be adapted to suit each particular case. Godson and Sinsheimer recommend, in difficult cases, warming for 1 or 2 min at 10-15°C after the lytic mixture is added. Completeness of cell lysis can be estimated by the loss of refractility of the cells under phase-contrast microscopy [85,86].

If desired, DNase and DOC can be omitted from the lytic mixture. DNA is then separated from the lysate by centrifugation at 3,000 X g for 5 min [85]. However, about 20-30% of the polysomes are trapped by the DNA. DNA is easily sedimented at 10 mM (or higher) Mg^2 concentration, but it becomes increasingly difficult to sediment at lower concentrations [85].

Extraction of polysomes is almost complete at low Mg^2 concentrations (5-10 mM). At 40 mM or higher almost no polysomes are extracted even though the cells release most of the soluble protein. DNase and DOC help in the extraction of polysomes at all Mg^2 concentrations [85]. If the extract is

1. The S-30 System from Escherichia coli

to be used for polypeptide synthesis, however, it may be advisable to lower the DOC concentration to 0.05% [68]. (See Chapter 5 for details.) Further extensive documentation on the EDTA-lysozyme extraction procedure, as well as its scaling up to lyse large amounts of cells, may be found in Refs. 85 and 86.

C. Polysome Extraction by Other Procedures

The EDTA-lysozyme method is used by several investigators with some modifications from the described procedure. Thus Flessel et al. [87] use chloramphenicol routinely and eliminate lysozyme and EDTA from the final lysate by separating the cells treated with these compounds by centrifugation before addition of the lytic mixture. Phillips et al. [72] prolong lysozyme-EDTA treatment to 5-8 min until formation of spheroplasts takes place. Spheroplasts are separated by centrifugation and lysed by Brij in the presence of DNase. Algranati et al. [76] also use chloramphenicol routinely, but lyse the cells with DOC in the presence of DNase without prior elimination of EDTA and lysozyme. All these procedures, as well as those of Ron et al. [84] and Godson and Sinsheimer [85] (previously detailed), appear to be capable of yielding extracts from exponentially growing cells with over 80% of the ribosomes in polysomes; they also extract 70 or 80% of the ribosomal RNA. Thus the choice of a particular procedure may be based on the experimenter's convenience, the characteristics of the E. coli strain, and the nature of the experiment.

A different methodology is described by Mangiarotti and Schlessinger [66]. Cells are cultivated in fragile form by growing them in the presence of 0.5 molal Na_2SO_4. Under these conditions cells grow at a slow but exponential rate and form chains of incompletely separated rods. They are collected by centrifugation and lysed by resuspension in a hypotonic medium containing DOC. About 70% of the cell RNA is released. If small amounts of penicillin are present during growth, fragility is increased and as much as 90 or 95% of the RNA is released. The main advantages of this method are the rapidity of lysis and the avoidance of lysozyme and EDTA treatment. The disadvantage, however, is that the growth in highly hypertonic medium and the osmotic shock during lysis may induce unwanted effects such as the dissociation of free 70 S ribosomes [74]. Mangiarotti and Schlessinger [66] also describe a method with fragile, sucrose-dependent mutants of E. coli. This method, although similar to the method of Kiho and Rich [89], is not generally applicable.

REFERENCES

[1] M. W. Nirenberg and J. H. Matthaei, Proc. Natl. Acad. Sci. U.S., 47, 1588 (1961).

[2] D. Nathans, G. Notani, J. H. Schwartz, and N. D. Zinder, Proc. Natl. Acad. Sci. U.S., 48, 1424 (1962).

1. The S-30 System from Escherichia coli   59

[3] D. Nathans, J. Mol. Biol., 13, 521 (1965).

[4] M. R. Capecchi, J. Mol. Biol., 21, 173 (1966).

[5] H. F. Lodish and H. D. Robertson, J. Mol. Biol., 45, 9 (1969).

[6] W. Salser, R. F. Gesteland, and A. Bolle, Nature, 215, 588 (1967).

[7] D. W. Slater and S. Spiegelman, Proc. Natl. Acad. Sci. U.S., 56, 164 (1966).

[8] K. A. Cammack and H. E. Wade, Biochem. J., 96, 671 (1965).

[9] A. Garen and O. Siddiqi, Proc. Natl. Acad. Sci. U.S., 48, 1121 (1962).

[10] E. S. Lennox, Virology, 1, 190 (1955).

[11] H. F. Lodish, J. Mol. Biol., 50, 689 (1970).

[12] J. H. Schwartz, Proc. Natl. Acad. Sci. U.S., 53, 1133 (1965).

[13] D. W. Slater and S. Spiegelman, Proc. Natl. Acad. Sci. U.S., 56, 164 (1966).

[14] R. E. Webster, D. L. Engelhardt, N. D. Zinder, and W. Konigsberg, J. Mol. Biol., 29, 27 (1967).

[15] M. W. Nirenberg, in Methods in Enzymology, Vol. 6 (S. P. Colowick and N. O. Kaplan, eds.), Academic Press, New York, 1963, p. 17.

[16] F. Lipmann, Science, 164, 1024 (1969).

[17] P. Lengyel and D. Söll, Bacteriol. Rev., 33, 264 (1969).

[18] J. M. Eisenstadt and G. Brawerman, Proc. Natl. Acad. Sci. U.S., 58, 1560 (1967).

[19] R. B. Roberts, P. H. Abelson, D. B. Cowie, E. T. Bolton, and R. J. Britten, Studies of Biosynthesis in E. coli, Publication 607, Carnegie Institution of Washington, Washington D.C., 1963, p. 420.

[20] W. Konigsberg, K. Weber, G. Notani, and N. Zinder, J. Biol. Chem., 241, 2579 (1966).

[21] M. R. Capecchi, Proc. Natl. Acad. Sci. U.S., 55, 1517 (1966).

[22] L. Luzzatto, D. Apirion, and D. Schlessinger, Proc. Natl. Acad. Sci. U.S., 60, 873 (1968).

[23] J. Modolell and B. D. Davis, Proc. Natl. Acad. Sci. U.S., 61, 1279 (1968).

[24] A. Tissières, D. Schlessinger, and F. Gros, Proc. Natl. Acad. Sci. U.S., 46, 1450 (1960).

[25] J. H. Matthaei and M. W. Nirenberg, Proc. Natl. Acad. Sci. U.S., 47, 1580 (1961).

[26] M. Revel and F. Gros, Biochem. Biophys. Res. Commun., 25, 124 (1966).

[27] G. Brawerman and J. M. Eisenstadt, Biochemistry, 5, 2784 (1966).

[28] M. Salas, M. B. Hille, J. A. Last, A. J. Wahba, and S. Ochoa, Proc. Natl. Acad. Sci. U.S., 57, 387 (1967).

[29] W. Gilbert, J. Mol. Biol., 6, 374 (1963).

[30] M. Revel and H. H. Hiatt, J. Mol. Biol., 11, 467 (1965).

[31] M. C. Ganoza and T. Nakamoto, Proc. Natl. Acad. Sci. U.S., 55, 162 (1966).

[32] W. Szer and S. Ochoa, J. Mol. Biol., **8**, 823 (1964).

[33] J. Davies, W. Gilbert, and L. Gorini, Proc. Natl. Acad. Sci. U.S., **51**, 883 (1964).

[34] M. R. Capecchi, J. Mol. Biol., **30**, 213 (1967).

[34a] D. L. Engelhardt, R. E. Webster, and N. D. Zinder, J. Mol. Biol., **29**, 45 (1967).

[35] H. F. Lodish, Nature, **220**, 345 (1968).

[36] P. Anderson, J. Davies, and B. D. Davis, J. Mol. Biol., **29**, 203 (1967).

[37] J. G. Flaks, E. C. Cox, and J. R. White, Biochem. Biophys. Res. Commun., **7**, 385 (1962).

[38] J. G. Speyer, P. Lengyel, and C. Basilio, Proc. Natl. Acad. Sci. U.S., **48**, 684 (1962).

[39] J. Davies, Proc. Natl. Acad. Sci. U.S., **51**, 659 (1964).

[40] B. Weisblum and J. Davies, Bacteriol. Rev., **32**, 493 (1968).

[41] D. Vazquez, Biochem. Biophys. Acta, **114**, 289 (1966).

[42] J. M. Wilhelm, N. L. Oleinick, and J. N. Corcoran, Antimicrobial Agents Chemotherapy-1967, p. 236 (1968).

[43] D. Vazquez, in Antibiotics, Vol. 1 (D. Gottlieb and P. D. Shaw, eds.), Springer Verlag, New York, 1967, p. 387.

[44] A. I. Laskin, in Antibiotics, Vol. 1 (D. Gottlieb and P. D. Shaw, eds.), Springer Verlag, New York, 1967, p. 331.

[45] B. E. H. Maden, R. R. Traut, and R. E. Monro, J. Mol. Biol., **35**, 333 (1968).

[46] M. A. Smith and M. A. Stahman, Biochem. Biophys. Res. Commun., 13, 251 (1963).

[47] K. Weber and W. Konigsberg, J. Biol. Chem., 242, 3563 (1967).

[48] M. R. Capecchi, Proc. Natl. Acad. Sci. U.S., 58, 1144 (1967).

[49] M. Sekiguchi and S. S. Cohen, J. Mol. Biol., 8, 638 (1964).

[50] M. Schweiger and L. M. Gold, Proc. Natl. Acad. Sci. U.S., 63, 1351 (1969).

[51] P. Albertson, in Methods in Virology, Vol. 2 (K. Maramorosch and H. Koprowki, eds.), Academic Press, New York, 1967, p. 312.

[52] R. F. Gesteland and H. Boedtker, J. Mol. Biol., 8, 496 (1964).

[53] J. Davies, L. Gorini, and B. D. Davis, Mol. Pharmacol., 1, 93 (1965).

[54] M. Sela and E. Katchalski, Advan. Protein Chem., 14, 391 (1959).

[55] R. S. Gardner, A. J. Wahba, C. Basilio, R. S. Miller, P. Lengyel, and J. F. Speyer, Proc. Natl. Acad. Sci. U.S., 48, 2027 (1962).

[56] G. R. Julian, J. Mol. Biol., 12, 9 (1965).

[57] A. J. Wahba, R. S. Gardner, C. Basilio, R. S. Miller, J. F. Speyer, and P. Lengyel, Proc. Natl. Acad. Sci. U.S., 49, 116 (1963).

1. The S-30 System from Escherichia coli 63

[58] M. W. Nirenberg, O. W. Jones, P. Leder, B. F. C. Clark, W. S. Sly and S. Pestka, Cold Spring Harbor Symp. Quant. Biol., 28, 549 (1963).

[59] J. F. Speyer, P. Lengyel, C. Basilio, A. J. Wahba, R. S. Gardner, and S. Ochoa, Cold Spring Harbor Symp. Quant. Biol., 28, 559 (1963).

[60] J. Davies, D. S. Jones, and H. G. Khorana, J. Mol. Biol., 18, 48 (1966).

[61] A. Tissières, J. D. Watson, D. Schlessinger, and B. R. Hollingworth, J. Mol. Biol., 1, 221 (1959).

[62] G. N. Godson, J. Mol. Biol., 34, 149 (1968).

[63] J. Modollel and B. D. Davis, Nature, 224, 345 (1969).

[64] J. Modollel and B. D. Davis, Proc. Natl. Acad. Sci. U.S., 67, 1148 (1970).

[65] R. E. Webster and N. D. Zinder, J. Mol. Biol., 42, 425 (1969).

[66] G. Mangiarotti and D. Schlessinger, J. Mol. Biol., 20, 123 (1966).

[67] R. Kaempfer, M. Meselson, and H. J. Raskas, J. Mol. Biol., 31, 277 (1968).

[68] R. Kaempfer, Proc. Natl. Acad. Sci. U.S., 61, 106 (1968).

[69] G. Mangiarotti and D. Schlessinger, J. Mol. Biol., 29, 395 (1967).

[70] C. Guthrie and M. Nomura, Nature, 219, 232 (1968).

[71] L. A. Phillips, B. Hotham-Iglewski, and R. M. Franklin, J. Mol. Biol., 40, 279 (1969).

[72] L. A. Phillips, B. Hotham-Iglewski, and R. M. Franklin, J. Mol. Biol., 45, 23 (1969).

[73] R. J. Beller and B. D. Davis, Bateriol. Proc., p. 49 (1970).

[74] R. E. Kohler, E. Z. Ron, and B. D. Davis, J. Mol. Biol., 36, 71 (1968).

[75] E. Z. Ron, R. E. Kohler, and B. D. Davis, J. Mol. Biol., 36, 83 (1968).

[76] I. D. Algranati, N. S. Gonzalez, and E. G. Bade, Proc. Natl. Acad. Sci. U.S., 62, 574 (1969).

[77] A. R. Subramanian, E. Z. Ron, and B. D. Davis, Proc. Natl. Acad. Sci. U.S., 61, 761 (1968).

[78] A. R. Subramanian, B. D. Davis, and R. J. Beller, Cold Spring Harbor Symp. Quant. Biol., 34, 223 (1969).

[79] E. G. Bade, N. S. Gonzalez, and I. D. Algranati, Proc. Natl. Acad. Sci. U.S., 64, 654 (1969).

[80] S. H. Miall, T. Kato, and T. Tamaoki, Nature, 226, 1050 (1970).

[81] J. Albrecht, F. Stap, H. O. Voorma, P. H. Van Knippenberg, and L. Bosch, FEBS Letters, 6, 297 (1970).

[82] R. Kaempfer and M. Meselson, Cold Spring Harbor Symp. Quant. Biol., 34, 209 (1969).

[83] R. Kaempfer, Nature, 228, 534 (1970).

1. The S-30 System from Escherichia coli

[84] E. Z. Ron, R. E. Kohler, and B. D. Davis, Science, 153, 1119 (1966).

[85] G. N. Godson and R. L. Sinsheimer, Biochim. Biophys. Acta, 149, 476, 489 (1967).

[86] G. N. Godson, in Methods in Enzymology, Vol. 12, Part A (L. Grossman and K. Moldave eds.), Academic Press, New York, 1967, p. 503.

[87] C. P. Flessel, P. Ralph, and A. Rich, Science, 158, 658 (1967).

[88] B. D. Davis and E. S. Mingioli, J. Bacteriol., 60, 17 (1950).

[89] Y. Kiho and A. Rich, Proc. Natl. Acad. Sci. U.S., 51, 111 (1964).

Chapter 2

Bacillus Subtilis PROTEIN-SYNTHESIZING SYSTEM

Roy H. Doi

Department of Biochemistry and Biophysics
University of California
Davis, California

I. INTRODUCTION 68

II. GROWTH OF B. subtilis VEGETATIVE CELLS AND SPORES . 70

    A. Growth of Vegetative Cells for Extracts Active in in vitro Protein Synthesis. 70

    B. Growth of Cells for Preparation of tRNA Fractions 71

    C. Growth of Spores for Spore Extracts 73

    D. Growth of Cells for the Preparation of Formyltetrahydrofolic Acid Synthetase 74

III. PREPARATION OF CELL-FREE EXTRACTS FOR PROTEIN SYNTHESIS. 75

    A. Vegetative Cell Extracts 75

    B. Spore Extracts 76

IV. ASSAY SYSTEMS FOR PROTEIN SYNTHESIS. 76

    A. Endogenous mRNA-Directed Synthesis of Polypeptides with Vegetative Cell Extracts 76

    B. Poly U-Directed Polyphenylalanine Synthesis by Vegetative Cell and Spore Extracts 81

V. CONCLUSIONS. 83

    REFERENCES 84

Copyright © 1971 by Marcel Dekker, Inc. No part of this work may be reproduced or utilized in any form or by any means, electronic or mechanical, including xerography, photocopying, microfilm, and recording, or by any information storage and retrieval system, without the written permission of the publisher.

I. INTRODUCTION

Protein synthesis has been examined with cell-free extracts from many species of Bacillus [1-17], and the conditions necessary for amino acid incorporation have been determined. In contrast to Escherichia coli, there has not been an extensive analysis of the components of the protein-synthesizing mechanism in Bacillus species, which are quite heterogenous in their physiological and genetic makeup. However, there have been relatively specific studies on tRNA fractions of B. subtilis during growth and sporulation [18-25], polymerization factors from Bacillus stearothermophilus [11,26-29], structure and properties of Bacillus ribosomes [30-44], reactions of Bacillus components with heterologous systems [5,45,46], effects of antibiotics on various reactions involved in polypeptide synthesis [46-49], and protein chain initiation [1,4,50,51].

The development of a good in vitro protein-synthesizing system with Bacillus extracts has been difficult, and the difficulties may be attributable to the inherent properties of these organisms. Bacilli produce numerous proteases and nucleases that are particularly abundant during the late stages of logarithmic growth and during sporulation. Lower concentrations of these enzymes can be found, moreover, at all stages of growth and can make the isolation of native macromolecular components difficult. Furthermore, many Bacillus species have a tendency to autolyze, particularly when harvested during the

2. Bacillus Subtilis Protein-Synthesizing System

log phase of growth, and the use of conditions to prevent premature lysis is required.

Bacillus species sporulate when the medium becomes deficient in glucose and/or nitrogen sources. During the sporulation stages the metabolism of the cells is altered, many physiological, enzymological, and structural changes occur, and an active vegetative cell is converted to a dormant spore. Spores can be readily and synchronously germinated under the proper conditions to produce active vegetative cells. This growth cycle should be clearly understood before experiments on macromolecular synthesis are undertaken with Bacillus species. The metabolic activities of the various growth stages are quite different and this is reflected in the heterogenous nature of cellular extracts, for example, in capacity for amino acid incorporation and in hydrolytic enzyme activities. Because of the special nature of these sporulating bacteria, one cannot arbitrarily "grow a culture overnight and harvest," but one must have cells from a well-defined stage of growth. In addition, dormant spores present a special problem since they are very difficult to lyse and do contain RNase and protease activities; special techniques have been developed for obtaining active spore lysates and undegraded molecules.

This chapter focuses on the cell-free protein-synthesizing systems developed for B. subtilis vegetative cells and spores. The system for vegetative cells is particularly effective for analyzing protein synthesis dependent on endogenous mRNA, since initiation and elongation occur readily, yielding polypeptides with N-formylmethionine or free amino acids at the NH_2 terminus and with molecular weights up to 40,000.

II. GROWTH OF B. subtilis VEGETATIVE CELLS AND SPORES

A. Growth of Vegetative Cells for Extracts Active in in vitro Protein Synthesis

Bacillus subtilis W23 cells are grown to early or midlog stage in Penassay (Difco) medium, or in modified Schaeffer's medium (SG medium) [52] containing per liter: nutrient broth (Difco), 16 g; $MgSO_4 \cdot 7H_2O$, 0.5 g; KCl, 2 g; glucose, 1 g; 10^{-3} M $Ca(NO_3)_2$, 10^{-4} M $MnCl_2$, and 10^{-6} M $FeSO_4$. The final four ingredients are each autoclaved separately and after cooling are added to their final concentrations.

The medium is inoculated to produce a density of approximately 2×10^7 cells per milliliter and shaken vigorously at 37°C. When the cell density reaches about $2-4 \times 10^8$ cells per milliliter, $MgCl_2$ is added to a concentration of 0.01 M to prevent lysis and the cells are chilled and harvested as quickly as possible by centrifugation at 4°C. The cells are washed once with 0.01 M tris-HCl buffer (pH 7.5) containing 0.01 M $MgCl_2$, 0.06 M KCl, 0.006 M mercaptoethanol, 0.5 mM spermidine, and 20% v/v glycerol (TMMG buffer). This buffer stabilizes the cell components necessary for protein synthesis. The washed

2. Bacillus Subtilis Protein-Synthesizing System

cells can be kept at $-20°C$ until used (at least 2 months). Cells harvested beyond the midlog phase of growth were progressively less active in in vitro protein synthesis.

B. Growth of Cells for Preparation of tRNA Fractions

Bacillus subtilis W23 cells are grown at $37°C$ with vigorous aeration to late log phase in Penassay (Difco) medium or in an enriched medium (TYG medium) containing 2.5% tryptone (Difco), 2.0% yeast extract (Difco), 0.3% K_2HPO_4, and 3% glucose at pH 7.5. With TYG medium it is possible to obtain 10-15 g (wet weight) of cells per liter. $MgCl_2$ is added to the culture to 0.01 M to prevent lysis and the cells are harvested by centrifugation at $4°C$.

tRNA is extracted by the phenol method [53] from <u>freshly</u> grown cells, since B. subtilis cells tend to lyse when frozen and thawed. If the tRNA is isolated from frozen and thawed cells, it is heavily contaminated with rRNA and with DNA. The cells are suspended in 2 vol of 0.01 M tris-HCl buffer (pH 7.4) containing 0.01 M acetate. An equal volume of phenol is added to the cell suspension and the mixture is shaken <u>gently</u> for 1 hr on a reciprocal shaker at room temperature. Vigorous shaking should be avoided since lysis and release of rRNA and DNA will occur. The mixture is centrifuged at 5,000 X g for 15 min. The upper aqueous phase containing the tRNA is removed and 1/2 vol of phenol is added, shaken for 15 min, and centrifuged as before. This extraction is repeated once more. To the final upper aqueous fraction, 1/10 vol of 20% (w/v) potassium acetate is added and the tRNA solution is cooled in an ice bath. 2 vol of cold 100% ethanol are added and the mixture is allowed to stand in an ice bath until precipitation occurs; with a high concentration of tRNA, precipitation is almost instantaneous. After the mixture is kept for several hours at 4°C, the precipitate is collected by centrifugation at 12,000 X g for 20 min.

The precipitate is dissolved in 0.5 M tris-HCl (pH 8.8) and incubated at 35°C for 1 hr to remove amino acids bound to the tRNA. Then the pH of the solution is carefully reduced to pH 7.0-7.5 with dilute HCl. In order to remove rRNA contaminating the preparation, 5 M NaCl is added to a final concentration of 1 M NaCl. The solution is kept in an ice bath for 1 hr and the precipitated rRNA is removed by centrifugation at 12,000 X g for 20 min.

The tRNA is precipitated from the resulting supernatant fraction by cooling the solution to 4°C and adding 2 vol of cold 100% ethanol. After 30 min the precipitated tRNA is collected by centrifugation at 12,000 X g for 15 min. The precipitate can be stored in suitable aliquots at -20°C or, if it is to be used, the tRNA can be dissolved in and dialyzed against 0.001 M tris-HCl buffer (pH 7.2). tRNA is more stable when frozen after ethanol precipitation than as a frozen

2. Bacillus Subtilis Protein-Synthesizing System 73

aqueous solution. If tRNA is to be stored for a long period, it is wise to add bentonite (of a size that can be sedimented by centrifugation at 12,000 X g for 10 min) to the RNA solution before the addition of ethanol. In this way the precipitated tRNA and bentonite can be sedimented together. Any RNase that may be contaminating the preparation is absorbed to the bentonite. When the tRNA is needed, the tRNA-bentonite precipitate is suspended in 0.001 M tris-HCl (pH 7.2), and the insoluble bentonite is removed from the tRNA solution by centrifugation.

C. Growth of Spores for Spore Extracts

Bacillus subtilis is grown in SG medium (described in Section I,A) with vigorous aeration at 37°C. Within 36-48 hr after inoculation, free spores should be observed. The spores are harvested by centrifugation and washed several times with water. Vegetative cells and germinated spores are separated from dormant spores by the method of Sacks and Alderton [54]. This technique uses a two-phase system employing polyethylene glycol and phosphate buffer. The polyethylene glycol-phosphate mixture is made by mixing 11.18 g of polyethylene glycol (Carbowax 4000) and 34.1 ml of 3 M phosphate buffer (pH 7.0), and adding the spore suspension plus water to make a final volume of 100 ml. This proportion of components can be used for any quantity of spores. The spore suspension is mixed vigorously and the mixture is left in a separatory funnel overnight. The bottom layer and the interface which contain vegetative cells and germinated spores, respectively, are removed, leaving a suspension of spores. The spores are collected by centrifugation and thoroughly washed with water; the procedure is then repeated with the polyethylene glycol-phosphate buffer until no vegetative cells or germinated spores are visible at the interface. The spores are then collected from the upper phase, thoroughly washed with distilled water, and kept either as a frozen pellet at $-20°C$ or lyophilized and kept at $-20°C$.

Since sporulation is difficult to obtain with large volumes of culture medium, usually because of limiting aeration, several 2.8-liter Fernbach flasks, each containing 1 liter of SG medium, can be used to obtain spores. One can use 10-liter Microferm fermenters at full aeration and obtain about 50-75% spore formation. In Fernbach flasks sporulation approaches 95-99%.

D. Growth of Cells for the Preparation of Formyltetrahydrofolic Acid Synthetase

In order to obtain maximum protein synthesis from the endogenous mRNA system, it is necessary to furnish formyltetrahydrofolic acid synthetase to the system. Crystalline formyltetrahydrofolic acid synthetase is purified from <u>Clostridium cylindrosporum</u> by the procedure described by

2. Bacillus Subtilis Protein-Synthesizing System

Rabinowitz and Pricer [55]. By starting with freshly grown cells, it is possible to obtain crystalline enzyme within 8 hr. Since this enzyme can be purified so readily, it is recommended that it be used for optimum results. If the enzyme is to be stored for a long period, it is better to use dithiothreitol (DTT) instead of mercaptoethanol in the storage buffer.

III. PREPARATION OF CELL-FREE EXTRACTS FOR PROTEIN SYNTHESIS

A. Vegetative Cell Extracts

Frozen cells are ground in a previously chilled mortar at 4°C with 4 mg of Macaloid and 2 vol of silica powder (150 mesh) per gram wet weight of cells. The enzymes are extracted with TMMG buffer equivalent to 2 times the volume of wet cells. The S-30 extract is obtained by centrifuging the crude extract twice at 30,000 X g for 20 min. The upper two-thirds of the final supernatant fraction is saved and dialyzed against TMMG buffer for 3 hr. This S-30 fraction is used immediately after dialysis for amino acid incorporation studies involving endogenous mRNA. No DNase is ever added to the extract. Endogenous mRNA-stimulated amino acid incorporation activity decays rapidly upon storage even when the extract is kept at -70°C. The extract can be stored for short periods (1 week) for studies utilizing poly U as the mRNA. However, the activity with endogenous mRNA was always better than with exogenous mRNA such as poly U or MS2 phage RNA.

B. Spore Extracts

Frozen or lyophilized spores are suspended first in modified TMMG buffer (with only 1.5% instead of 20% glycerol) and then centrifuged at 10,000 X g for 20 min to form a pellet. This spore pellet is usually loose, thus requiring careful removal of the buffer. The spore pellet is frozen and thawed three times in liquid nitrogen to weaken the tough spore coats in the absence of germination. The spores are never allowed to warm up during the thawing step so that the temperature is maintained at around 0-5°C. After the spores are frozen for the fourth time, the frozen pellet is ground in a prechilled mortar with fragments of dry ice. The spores are usually weakened enough by the freezing and thawing treatment to give 90-95% breakage. A volume of regular TMMG buffer is added to the broken spore paste, equivalent to the volume of the original spore pellet. The extract is centrifuged at 17,000 X g for 15 min and the supernatant fraction is centrifuged twice at 30,000 X g for 20 min. The upper two-thirds of the supernatant fraction is removed and dialyzed against TMMG buffer for 3 hr. This preparation is not stable and should be used immediately after dialysis; it is referred to as spore S-30 fraction.

IV. ASSAY SYSTEMS FOR PROTEIN SYNTHESIS

A. Endogenous mRNA-Directed Synthesis of Polypeptides with Vegetative Cell Extracts

2. Bacillus Subtilis Protein-Synthesizing System

Component	Amount
Tris-HCl buffer (pH 7.5)	50 µmoles
$MgCl_2$	4 µmoles
Ammonium acetate	20 µmoles
ATP	1 µmole
GTP	0.005 µmole
Mercaptoethanol	10 µmoles
Sodium formate (unlabeled or ^{14}C-labeled)	0.1 µmole
19 Unlabeled amino acids	0.03 µmole (each)
^{14}C- or ^{3}H-labeled amino acid (1 µCi)	0.02 µmole
Phosphoenolpyruvate	2.5 µmoles
Phosphoenolpyruvate kinase	10 µg
Tetrahydrofolic acid	50 µg
Formyltetrahydrofolic acid synthetase	40 units [55]
tRNA (B. subtilis)	400-600 µg
S-30 Protein	1-2 mg

The reaction mixture shown in the accompanying table can be used to analyze the initiation and synthesis of large polypeptides. It allows sufficient incorporation of amino acids for the analysis of newly made peptides for N-formylmethionine, N-formylmethionylalanine, free NH_2-terminal residues, and their molecular size. The reaction mixture is contained in a total volume of 0.5 ml.

The reaction mixture (minus the S-30 enzymes) is first incubated for 5 min at 37°C to form formyl-H_4-folate; then the S-30 enzyme fraction is added and the reaction mixture incubated for the desired time at 37°C.

If formate incorporation into N-formylmethionine is to be studied, [^{14}C]formate (sodium [^{14}C]formate, 40 Ci/mole) and [^{3}H]methionine (5 Ci/mmole) can be used to double-label the product. Unlabeled formate can be used for studying amino acid incorporation for other purposes. Since the initial rate of synthesis and the total incorporation of amino acids is at least twice as great with the formyl-H_4-folate-synthesizing system, the inclusion of this system is highly recommended.

The reaction is stopped and assayed in various ways depending on the purpose of the study:

(1) For studying the general incorporation of labeled amino acids into polypeptides, the reaction is stopped at the desired time by the addition of 1 ml of cold 10% trichloroacetic acid (TCA) containing 2 mg of the unlabeled form of the amino acid used to label the product in the reaction mixture. Then the reaction mixture is heated to 95°C for 20 min, cooled in an ice bath, filtered (type AP20 glass-fiber filter, Millipore Corporation, Bedford, Massachusetts), and the precipitate is washed five times with cold 10% TCA, once with 10% TCA containing 2 mg of unlabeled amino acid, and finally with 10 ml of cold

2. Bacillus Subtilis Protein-Synthesizing System

100% ethanol. The filter is dried and the radioactivity counted in a liquid scintillation counter.

For kinetic studies a series of identical reaction mixtures is prepared and stopped with 1 ml of cold TCA at the desired times, or 0.2-ml aliquots are removed at the desired time from a severalfold larger reaction mixture and added to 1 ml of cold TCA to stop the reaction.

(2) For isolation of N-formylmethionine from newly made polypeptides, the reaction mixture containing [^{14}C]formate and [^{3}H]methionine is rapidly chilled in ice and 50 µg of pancreatic RNase and 25 µg of DNase are added. After 5 min at 0°C to allow DNase action, the reaction mixture is made up to 0.02 M EDTA to chelate Mg^2 and to allow hydrolysis of N-formylmethionyl-tRNA by RNase for another 15 min at 0°C. Any remaining aminoacyl-tRNA is cleaved by treating the mixture with 0.1 M KOH (pH 12) for 20 min at 0°C. The proteins are precipitated with 3 vol of 10% TCA; the precipitate is then washed three times with 10% TCA. After removal of the TCA with ethanol-ether (3:1) washes, the protein is dried under nitrogen. The dried residue is suspended in 1 ml of 2% NH_4HCO_3 containing 0.001 M sodium thioglycolate. Pronase (0.5 mg) is added and the hydrolysis carried out at room temperature for 16 hr. The hydrolysis is terminated by acidifying the reaction mixture with concentrated HCl. After the precipitate is removed by centrifugation, the supernatant is extracted with ethyl acetate

to remove the formylated amino acids [56]. The extract is dried under nitrogen and subjected to high-voltage paper electrophoresis in pyridine acetate (pH 6.4) at 2 kV for 1 hr. Whatman no. 1 (or no. 3MM) paper is used. After drying, the paper is cut into 1-cm strips and counted in a liquid scintillation counter.

(3) For the isolation of N-formylmethionylalanine, the procedure described above is followed with minor modifications. The ethanol-ether-washed TCA-precipitated proteins are suspended in 1 ml of 2% NH_4CO_3. At this point 60 µg of unlabeled carrier N-formylmethionylalanine is added to the protein. Pronase (0.5 mg) is added and the hydrolysis carried out at 0°C for 12 hr. The Pronase digest is analyzed as described above.

(4) The procedure of Fraenkel-Conrat et al. [57] is used for the analysis of free NH_2-terminal residues of newly made peptides. The reaction mixture containing the appropriate ^{14}C- or ^3H-labeled amino acid is stopped quickly by chilling the reaction mixture in an ice bath. The aminoacyl- and peptidyl-tRNA bonds are cleaved by incubating the reaction mixture at pH 12 for 20 min at 0°C. The proteins are precipitated with cold 10% TCA and washed three times with cold 10% TCA. The TCA is removed by washing the protein with ethanol-ether (3:1). The reaction product with fluorodinitrobenzene is then washed successively with HCl, ether, and acetone. DNP-

2. Bacillus Subtilis Protein-Synthesizing System

proteins are hydrolyzed in an evacuated sealed tube with constant-boiling HCl (0.5 ml) for 18 hr at 105°C. DNP-amino acids are extracted with ether from the hydrolyzate. The radioactivity in the ether-soluble fraction is then measured by liquid scintillation counting.

(5) For determination of the size of the product formed in vitro, a reaction mixture of 4 times the standard volume is quickly chilled in ice and incubated with RNase (50 µg) and DNase (25 µg) for 5 min. The mixture is then made up to 0.02 M EDTA and incubated for 15 min at 0°C. The reaction mixture is then treated at pH 12 at 0°C for 20 min to cleave peptidyl-tRNA bonds. The larger proteins and peptides are precipitated by the addition of cold 2 N HCl. The precipitate is washed three times with cold dilute HCl. The precipitate is then dissolved in 1 ml of 8 M urea containing 0.1 M NH_3 and passed through a Sephadex G-100 column (0.9 X 132 cm) at a flow rate of 5.7 ml/hr. 2-ml fractions are collected. The column is calibrated for the determination of molecular weight by passing blue dextran, bovine serum albumin, cytochrome c, insulin, and tryptophan through it under identical conditions.

B. Poly U-Directed Polyphenylalanine Synthesis by Vegetative Cell and Spore Extracts

Since spores contain little or no mRNA, it is necessary to use poly U as mRNA to obtain amino acid incorporation. For both vegetative cells and spore extracts, the rate and extent of amino acid incorporation is relatively poor (10%) compared to that obtained with E. coli extracts. Also, little or no incorporation is observed when MS2 phage RNA is used as messenger; in fact, MS2 RNA inhibits the endogenous mRNA-directed protein synthesis of vegetative cells. The reaction mixture shown in the accompanying table can be used to analyze both

Component	Amount
Tris-HCl buffer (pH 7.5)	50 µmoles
$MgCl_2$	10 µmoles
Ammonium acetate	20 µmoles
ATP	1 µmole
GTP	0.005 µmole
Mercaptoethanol	10 µmoles
19 Unlabeled amino acids	0.03 µmole each
^{14}C- or ^{3}H-labeled phenylalanine (1 µCi)	0.03 µmole
Spermidine	0.125 µmole
Phosphoenolpyruvate	2.5 µmoles
Phosphoenolpyruvate kinase	10 µg
Poly U	50-250 µg
tRNA (B. subtilis)	500 µg
S-30 Proteins	1-2 mg

vegetative cell and spore extracts for poly U-directed protein synthesis and contains the indicated components per 0.5 ml.

2. Bacillus Subtilis Protein-Synthesizing System

The reaction mixture, minus poly U and labeled phenylalanine, is always preincubated for 10 min at 37°C; then the poly U and labeled phenylalanine are added to start the incorporation of labeled phenylalanine. After incubation at 37°C for the desired time, the reaction is stopped by adding 1 ml of 10% TCA containing 2 mg of unlabeled phenylalanine. The reaction mixture is heated to 95°C for 20 min, cooled in an ice bath, filtered, and the precipitate washed five times with cold 10% TCA, once with 10% TCA containing 2 mg of phenylalanine, and finally with 10 ml of cold 100% ethanol. The filters are dried and counted.

V. CONCLUSIONS

The requirements for in vitro protein synthesis by B. subtilis extracts are similar to those reported for E. coli. However, the use of the specific conditions described in this chapter for the growth, harvesting, and breakage of cells, and the use of endogenous mRNA allows protein synthesis in vitro that is very similar to that expected in vivo. With this system initiation and synthesis of large polypeptides is possible. However, the use of exogenous mRNA, such as MS2 phage RNA or poly U, has not been highly successful; this suggests some degree of specificity of the system for homologous mRNA. It has also been difficult to perform reconstitution experiments involving supernatant fractions and ribosomes of B. subtilis, since isolated ribosomes are extremely labile [5]. However, reasonably good poly U-directed activity with isolated ribosomes has been obtained with Bacillus cereus [6], B. megaterium [8], and B. licheniformis [15] ribosomes. The use of RNase-deficient mutants of B. subtilis may overcome some of the difficulties encountered with B. subtilis extracts. In any case relatively crude systems with homologous mRNA fractions can be used to acquire useful information not obtainable by other means.

REFERENCES

[1] L. K. Migita and R. H. Doi, J. Biol. Chem., 245, 2005 (1970).

[2] H. Bishop, L. K. Migita, and R. H. Doi, J. Bacteriol., 99, 771 (1969).

[3] A. Hirashima, K. Asano, and A. Tsugita, Biochim. Biophys. Acta, 134, 165 (1967).

[4] M. Takeda and R. E. Webster, Proc. Natl. Acad. Sci. U.S., 60, 1487 (1968).

[5] M. Takeda and F. Lipmann, Proc. Natl. Acad. Sci. U.S., 56, 1875 (1966).

[6] Y. Kobayashi and H. O. Halvorson, Biochim. Biophys. Acta, 119, 160 (1966).

[7] A. G. Atherly and J. Imsande, Biochim. Biophys. Acta, 145, 491 (1967).

[8] M. P. Deutscher, P. Chambon, and A. Kornberg, J. Biol. Chem., 243, 5117 (1968).

2. Bacillus Subtilis Protein-Synthesizing System

[9] D. Schlessinger, J. Mol. Biol., 7, 569 (1963).

[10] B. Bubela and E. S. Holdsworth, Biochim. Biophys. Acta, 123, 376 (1966).

[11] J. Waterson, G. Beaud, and P. Lengyel, Nature, 227, 34 (1970).

[12] S. M. Friedman and I. B. Weinstein, Biochim. Biophys. Acta, 114, 593 (1966).

[13] N. S. Gonzalez, S. H. Goldemberg, and I. D. Algranati, Biochim. Biophys. Acta, 166, 760 (1968).

[14] J. W. Davies, Biochim. Biophys. Acta, 174, 686 (1969).

[15] J. Stenesh and N. Schechter, J. Bacteriol., 98, 1258 (1969).

[16] L. M. Changchien and J. N. Aronson, J. Bacteriol., 103, 734 (1970).

[17] G. Coleman, Biochim. Biophys. Acta, 182, 180 (1969).

[18] I. Kaneko and R. H. Doi, Proc. Natl. Acad. Sci. U.S., 55, 564 (1966).

[19] R. H. Doi, I. Kaneko and R. T. Igarashi, J. Biol. Chem., 243, 947 (1968).

[20] R. H. Doi, I. Kaneko, and B. Goehler, Proc. Natl. Acad. Sci. U.S., 56, 1548 (1966).

[21] J. L. Arceneaux and N. Sueoka, J. Biol. Chem., 244, 5959 (1969).

[22] M. P. Stulberg, K. R. Isham, and A. Stevens, Biochim. Biophys. Acta, 186, 297 (1969).

[23] R. A. Lazzarini, Proc. Natl. Acad. Sci. U.S., 56, 185 (1966).

[24] R. A. Lazzarini and E. Santangelo, J. Bacteriol., 94, 125 (1967).

[25] B. Vold, J. Bacteriol., 102, 711 (1970).

[26] A. Skoultchi, Y. Ono, H. M. Moon, and P. Lengyel, Proc. Natl. Acad. Sci. U.S., 60, 675 (1968).

[27] Y. Ono, A. Skoultchi, J. Waterson, and P. Lengyel, Nature, 222, 645 (1969).

[28] Y. Ono, A. Skoultchi, J. Waterson, and P. Lengyel, Nature, 223, 697 (1969).

[29] A. Skoultchi, Y. Ono, J. Waterson, and P. Lengyel, Biochemistry, 9, 508 (1970).

[30] R. H. Doi and R. T. Igarashi, J. Bacteriol., 87, 323 (1964).

[31] H. L. Bishop and R. H. Doi, J. Bacteriol., 91, 695 (1966).

[32] P. Morell, I. Smith, D. Dubnau, and J. Marmur, Biochemistry, 6, 258 (1967).

[33] M. Bleyman and C. Woese, J. Bacteriol., 97, 27 (1969).

[34] R. M. Pfister and D. G. Lundgren, J. Bacteriol., 88, 1119 (1964).

[35] P. C. Fitz-James, Can. J. Microbiol., 10, 92 (1964).

[36] F. M. Feinsod and H. A. Douthit, Science, 168, 991 (1970).

[37] A. Aronson, J. Mol. Biol., 15, 505 (1966).

2. Bacillus Subtilis Protein-Synthesizing System

[38] J. M. Idriss and H. O. Halvorson, Arch. Biochem. Biophys., 133, 442 (1969).

[39] P. Chambon, M. P. Deutscher, and A. Kornberg, J. Biol. Chem., 243, 5110 (1968).

[40] E. Otaka, T. Itoh, and S. Osawa, J. Mol. Biol., 33, 93 (1968).

[41] N. Nanninga, J. Mol. Biol., 48, 367 (1970).

[42] L. Stevens, Biochem. J., 113, 117 (1969).

[43] M. Takai and M. Kondo, Biochim. Biophys. Acta, 55, 875 (1962).

[44] G. F. Saunders and L. L. Campbell, J. Bacteriol., 91, 332 (1966).

[45] S. Lederberg and V. Lederberg, Exptl. Cell Res., 25, 198 (1961).

[46] F. N. Chang, C. J. Sih, and B. Weisblum, Proc. Natl. Acad. Sci. U.S., 55, 431 (1966).

[47] N. L. Oleinick, J. M. Wilhelm, and J. W. Corcoran, Biochim. Biophys. Acta, 155, 290 (1968).

[48] A. Ahmed, Biochim. Biophys. Acta, 166, 218 (1968).

[49] E. Cundliffe, Mol. Pharmacol., 3, 401 (1967).

[50] K. Horikoshi and R. H. Doi, Arch. Biochem. Biophys., 122, 685 (1967).

[51] L. K. Migita and R. H. Doi, Arch. Biochem. Biophys., 138, 457 (1970).

[52] P. Schaeffer, H. Ionesco, A. Ryter, and G. Balassa, Colloq. Intern. Centre Natl. Rech. Sci. (Paris), No. 124, 553 (1963).

[53] G. von Ehrenstein and F. Lipmann, Proc. Natl. Acad. Sci. U.S., 47, 941 (1961).

[54] L. E. Sacks and G. Alderton, J. Bacteriol., 82, 331 (1961).

[55] J. C. Rabinowitz and W. E. Pricer, Jr., in Methods in Enzymology, Vol. 6 (S. P. Colowick and N. O. Kaplan, eds.), Academic Press, New York, 1963, p. 375.

[56] P. Leder and H. Burtsztyn, Proc. Natl. Acad. Sci. U.S., 56, 1579 (1966).

[57] H. Fraenkel-Conrat, J. I. Harris, and A. L. Levy, Methods Biochem. Anal., 2, 359 (1955).

Chapter 3

PROTEIN SYNTHESIS SYSTEMS FROM HALOPHILIC BACTERIA

S. T. Bayley

Department of Biology
McMaster University
Hamilton, Ontario
Canada

I. INTRODUCTION 89

II. GROWTH AND HARVESTING OF BACTERIA. 91

III. HOMOGENIZATION OF CELLS AND PREPARATION OF S-60 EXTRACT. 96

IV. ASSAY OF S-60 EXTRACT. 98

V. PREPARATION OF RIBOSOMES FREE OF mRNA AND OF S-150 EXTRACTS 101

VI. INCORPORATION WITH SYNTHETIC mRNAs 103

VII. PREPARATION OF AMINOACYL-tRNA SYNTHETASES AND tRNA . 104

VIII. MEASUREMENT OF AMINOACYL-tRNA FORMATION. 107

REFERENCES 109

I. INTRODUCTION

An organism is described as halophilic if it grows best in concentrations of NaCl higher than the normal physiological range. This chapter is concerned with preparations from a species of <u>Halobacterium</u>, the most halophilic of all bacteria. Halobacteria grow best in NaCl concentrations between 25% and saturation and in fact high concentrations of NaCl are essential for their survival; at concentrations below about 3% NaCl, they lyse (for reviews see Refs. 1-3).

Copyright © 1971 by Marcel Dekker, Inc. No part of this work may be reproduced or utilized in any form or by any means, electronic or mechanical, including xerography, photocopying, microfilm, and recording, or by any information storage and retrieval system, without the written permission of the publisher.

Of immediate interest is the fact that the total concentration of salt within halobacteria is comparable to that in the medium. Direct measurements by Christian and Waltho [4] showed that <u>Halobacterium salinarium</u> grown in 4 M NaCl and 0.03 M KCl contained concentrations of KCl and NaCl greater than 4 M and 1 M, respectively. Thus the metabolic processes of these cells are carried out in essentially saturated salt, that is, under conditions in which ionic interactions between macromolecules must be minimized. In vitro systems from these cells therefore afford an opportunity for examining the relative importance of this type of weak interaction in processes such as translation.

Studies on <u>Halobacterium cutirubrum</u> [5,6] have shown that in its essentials the mechanism of translation is the same as in nonhalophilic bacteria. The main difference is in the concentration of salts that the translation machinery requires for activity. Studies on the fidelity of translation in the halophile system [7,8] suggest that the critical recognition processes on which fidelity depends, namely, attachment of amino acids to specific tRNAs and codon-anticodon matching, are unchanged but that it is the ancillary machinery, that is, the enzymes and many of the ribosomal proteins, that has adapted to an environment of concentrated salt. In the following discussion, therefore, the procedure is identical in outline to that for preparing protein-synthesizing systems and components from, for example, <u>Escherichia coli</u>, except that concentrated salts

3. Protein Synthesis from Halophilic Bacteria

(principally KCl) are essential at all stages to preserve the functional integrity of ribosomes and enzymes.

II. GROWTH AND HARVESTING OF BACTERIA

As experimental material, halobacteria have some advantages that compensate for the problems created by the high concentrations of salt in their media. Because of their peculiar salt requirements, they are very unlikely to be pathogenic [9], and contamination of cultures by other common laboratory microorganisms is negligible. Furthermore, because halobacteria lyse in low concentrations of salt, they can be flushed away with water.

Studies on protein synthesis have been carried out entirely on H. cutirubrum strain 9 obtained from the Culture Collection at the National Research Council of Canada, Ottawa, Canada. Although other Halobacterium spp. have not been examined, their growth characteristics and the ease with which they lyse are sufficiently similar to the properties of H. cutirubrum for there to be no reason to expect any serious problem in applying the procedures described here to other species.

The growth and culturing of Halobacterium, as well as of other halophilic bacteria, has been recently reviewed by Gibbons [9]. A satisfactory complex medium is shown in Table 1. This medium has been used throughout the work in our laboratory on cell-free protein synthesis. A synthetic medium [11,12], which

TABLE 1

Complex Medium for Extremely Halophilic Bacteria[a,b]

Component	Amount (g)
Casamino acids (Difco)	7.5
Yeast extract (Difco)	10
Sodium citrate	3
KCl	2
$MgSO_4 \cdot 7H_2O$	20
NaCl	250

[a]A solution of 5% $FeSO_4 \cdot 7H_2O$ is acidified with 1 ml of 1 N HCl per 100 ml. To about 800 ml of water, add with stirring the casamino acids and yeast extract, followed by the salts and 1 ml of $FeSO_4$ solution. Adjust the pH to 7.4-7.6 with NaOH and autoclave for 5 min. Cool, filter, make the filtrate to 1 liter with water, and adjust the pH to 6.2 with HCl. Autoclave for 20 min before use.

[b]This medium, based on the work of Sehgal and Gibbons [10], contains 4.0 M NaCl, 0.03 M KCl, 0.1 M $MgSO_4$, and 10 ppm Fe^{2+}.

gives growth comparable to that with the complex medium but avoids the tedious filtering necessary with the latter, is presented in Table 2.

Cells are grown at 37°C. Halobacteria are obligate aerobes so that aeration is essential. We routinely grow batch cultures

3. Protein Synthesis from Halophilic Bacteria

TABLE 2

Synthetic Medium for Extremely Halophilic Bacteria[a,b]

Fifteen amino acids:

DL-Alanine, 0.43 g; L-arginine, 0.40 g; L-cystine, 0.05 g;
L-glutamic acid, 1.30 g (or DL-aspartic acid, 0.45 g);
glycine, 0.06 g; DL-isoleucine, 0.44 g; L-leucine, 0.80 g;
L-lysine, 0.85 g; DL-methionine, 0.37 g; DL-phenylalanine,
0.26 g; L-proline, 0.05 g; DL-serine, 0.61 g; DL-threonine,
0.50 g; L-tyrosine, 0.20, DL-valine, 1.00

Two nucleotides:

Poly A, 0.10 g; poly U, 0.10 g

Glycerol, 1.0 g

Salts:

NaCl, 250 g; $MgSO_4 \cdot 7H_2O$, 20 g; NH_4Cl, 5 g; KCl, 2 g;
KNO_3, 100 mg; K_2HPO_4, 50 mg; KH_2PO_4, 50 mg; sodium citrate,
500 mg; $CaCl_2 \cdot 7H_2O$, 7.0 mg; $FeCl_2$, 2.30 mg; $ZnSO_4 \cdot 7H_2O$,
440 μg; $MnSO_4 \cdot H_2O$, 300 μg; $CuSO_4 \cdot 5H_2O$, 50 μg

---

[a]Adjust pH of medium to 6.2 with KOH and make the final volume up to 1 liter with distilled water.

[b]Method of Onishi et al. [11] modified by the addition of KCl by Gochnauer and Kushner [12].

of up to 12 liters each in a New Brunswick Microferm bench fermentor with an airflow of 3 liters/min and a stirring speed of 300 rpm. Cultures of 100 liters have been grown satisfactorily in 150-liter fermentors, although in this case oxygen was introduced with the air. Cultures of up to 5 liters each can be grown satisfactorily in baffled glass flasks placed on a rotary shaker and aerated through two or more sintered glass spargers. In all cases foaming is prevented by adding Dow Corning Antifoam A antifoaming agent.

Batch cultures are inoculated with a 5% inoculum. This starter culture is grown for 24-30 hr in an Erlenmeyer flask on a rotary shaker. For inoculating starter cultures it is convenient to maintain cells in 100-ml cultures in Erlenmeyer flasks on a rotary shaker. These can be subcultured weekly for 3-4 months, after which time they usually deteriorate and fresh cultures must be started from agar slants.

Under good growth conditions H. cutirubrum has a generation time of 6-8 hr. To obtain active cell-free preparations, it is essential to harvest actively growing cultures in early log phase, that is, after 18-24 hr, to yield between 0.6 and 1.2 g wet weight of cells per liter of culture. Unless otherwise mentioned, harvesting and all subsequent procedures are carried out at 0-4°C.

Cultures can be harvested by centrifuging at about 8000 X g for 10 min (7000 rpm in a Servall GSA rotor is suitable). However, harvesting, particularly of low-density cultures, is most conveniently carried out in a centrifuge of the Sharples

3. Protein Synthesis from Halophilic Bacteria

type running at 30,000-35,000 rpm. A convenient method of recovering small quantities of cells from a centrifuge is to cut a sheet of photographic film (unexposed but acid-fixed and washed) so as to form a liner for the rotor. This can easily be removed after the centrifuge run and the cells can be scraped from it.

It is probably well to point out that concentrated salt solutions badly corrode metals with which they are allowed to remain in contact for any length of time. It pays to specify the finest quality stainless steel for any equipment, such as fermentor assemblies and rotors for Sharples-type centrifuges, that comes into direct contact with the solutions. All metal parts should be thoroughly washed after use, and this applies particularly to aluminum alloy caps and rotors for preparative ultracentrifuges. Laxity on this score can be expensive in regard to equipment and to the friendship of laboratory colleagues.

To remove traces of medium, the harvested cells are dispersed in a convenient volume (25 ml or more per gram wet weight of cells) of the wash solution shown in Table 3, and pelleted by centrifuging at 8000 X g for 10-15 min (e.g., 7000 rpm in a Servall GSA rotor). If cells are to be stored frozen, this is the stage at which to store them, but because of the high intracellular salt concentrations temperatures below about $-40°C$ must be used. However, in our laboratory fresh cells are

TABLE 3

Solution for Washing Cells[a]

Component	Amount (g)
NaCl	500
$MgSO_4 \cdot 7H_2O$	40
KCl	4

[a]Make up to 2 liters with water.

always used for preparing cell-free systems for protein synthesis to avoid any possible losses of activity during storage.

III. HOMOGENIZATION OF CELLS AND PREPARATION OF S-60 EXTRACT

Cells are broken by taking advantage of the fact that although they are stable in concentrated NaCl they lyse readily in KCl solutions. Centrifugation from the wash solution should be sufficient to give a firm pellet of cells from which the supernatant can be thoroughly drained. The cells are then dispersed, in a volume corresponding to 5-10 ml per gram wet weight of cells, in a working buffer called solution D and consisting of 3.4 M KCl, 0.1 M magnesium acetate, and 0.01 M tris-HCl (pH 7.6); the suspension is centrifuged at 23,000 X g for 15 min (e.g., 14,000 rpm in a Servall SS-34 rotor). In solution D some of the cells lyse so that the upper part of the pellet consists of loosely packed, viscous, red material.

3. Protein Synthesis from Halophilic Bacteria

As much supernatant as possible is poured off and the tubes are weighed to estimate roughly the wet weight of cells.

The procedure for preparing and assaying cell-free extracts is based on that developed by Nirenberg [13] for E. coli. The cells are stirred with a volume of solution D corresponding to about 1.5 times their wet weight, together with 0.015 ml of β-mercaptoethanol and 1 mg of electrophoretically purified DNase (Worthington Biochemical Corporation) per 30 ml of solution D. The DNase is added as a solution of 1 mg/ml in 0.1 M potassium acetate (pH 5). (Other reducing agents would probably work as satisfactorily as β-mercaptoethanol; any DNase preparation used must of course be free of RNase.) The cells are then homogenized in a glass-Teflon Potter-Elvehjem homogenizer. Four to six passes of the Teflon plunger should be sufficient to break most of the cells. During this time the homogenizer should be kept cold and since the homogenate becomes very viscous care should be taken in case the glass tube breaks under pressure, as can happen if it becomes scratched with use.

To remove unbroken cells and fragments of cell envelopes, the homogenate is centrifuged once at 40,000 X g for 20 min (25,000 rpm in a no. 40 or no. 65 Beckman rotor) and twice at 60,000 X g for 30 min (30,000 rpm). In each case only the upper, clear part of the supernatant is retained. The final supernatant is then dialyzed for 4 hr against three changes of an excess of solution D containing 1 μl of β-mercaptoethanol per 2 ml of solution D as before. The final product constitutes S-60 extract.

S-60 extract can be kept for long periods without loss of activity by freezing in isopentane cooled in liquid nitrogen and storing under liquid nitrogen. A convenient method is to seal the extract in 1- to 1.5-ml portions in annealed thick-walled glass vials (10-mm outer diameter 2-mm wall is suitable). After freezing, the vials are suspended by strings in liquid nitrogen inside a large Dewar flask. To prevent the string from pulling off the vial, an inverted glass hook must be added to the side of the vial toward its upper end, below which the string is tied. Unless the vials are well annealed, they may crack, fill with liquid nitrogen, and explode on thawing. In handling frozen vials, therefore, it is advisable to wear goggles or a face mask and to allow vials to thaw inside a metal can.

IV. ASSAY OF S-60 EXTRACT

The A_{260} of S-60 extract varies among preparations. It should be 250 or more, although extracts with values as low as 100 can be used. The protein content of extracts can be estimated by the procedure of Lowry et al. [14]. The ribosomal content can be estimated by measuring the decrease in A_{260} after centrifuging the extract at 150,000 X g for 2 hr, and from this decrease calculating the equivalent concentration of ribosomes

3. Protein Synthesis from Halophilic Bacteria

using $E^{1\%}_{1\,cm}$ at 260 nm = 158 [15]. Extracts with A_{260} = 250 contain about 30 mg protein and about 5 mg or more of ribosomes per milliliter.

To assay for the incorporation of amino acids, the reaction mixture of nominal volume 0.125 ml contains the following components in 3.8 M KCl, 1.0 M NaCl, 0.4 M NH_4Cl, 0.04 M magnesium acetate, and 0.03 M tris-HCl buffer (pH 8): sodium ATP, 0.15 μmole; sodium phosphoenolpyruvate (PEP), 0.6 μmole; sodium or lithium GTP, 0.12 μmole; a mixture of 19 amino acids omitting the ^{14}C-labeled amino acid, 0.004 μmole of each; [^{14}C]amino acid (0.25 μCi); and S-60 fraction, 0.6 mg of protein (0.12 mg ribosomes). To obtain a concentration of 3.8 M KCl in this mixture, it is necessary to add dry salt. In making up the mixture, no allowance is made for the volume of this salt, so the final concentrations of salts are lower than the values quoted above. The exact composition of the reaction mixture is given in Table 4. Other components, for example, RNase to provide a control (about 10 μg per reaction mixture), can be added in aqueous solution in place of water.

In making up the reaction mixture, it is convenient to weigh the dry KCl into the incubation tube, then add the solutions, ending with the labeled amino acid and S-60 extract in that order. As much as possible of the KCl is dissolved by stirring with a glass rod. The remainder dissolves rapidly during incubation. As in assaying other cell-free systems,

TABLE 4

Reaction Mixture for Incorporation of Amino

Acids by S-60 Extract

Component	Amount
0.03 M Sodium ATP	5 μl
0.12 M Sodium PEP	5 μl
0.012 M Sodium GTP or lithium GTP	10 μl
Cold amino acid mixture less labeled amino acid (each 8 X 10^{-4} M	5 μl
3.4 M KCl, 0.12 M magnesium acetate, 0.15 M tris (pH 8.05)	25 μl
1.25 M NH_4Cl, 3.125 M NaCl	40 μl
Solid KCl	24 mg
^{14}C-Amino acid (neutralized)	5 μl
S-60 Fraction in solution D	20 μl
Water	10 μl
Total nominal volume[a]	125 μl

[a]The final salt concentrations are calculated to be 3.8 M KCl, 1.0 M NaCl, 0.4 M NH_4Cl, 0.04 M magnesium acetate, and 0.3 M tris-HCl buffer on the basis of the total volume of solutions added. Components in the mixture can be altered provided that these calculated concentrations remain the same. Because no allowance is made for the volume of dry salt added, the final concentrations of salts are actually lower than these.

3. Protein Synthesis from Halophilic Bacteria

frozen aliquots of previously prepared ATP, GTP, and PEP solutions and of S-60 extract are thawed for making up the reaction mixture and the remainder is discarded.

The reaction mixture is incubated at 37°C for 40 min; incorporation is stopped by cooling to 0°C and adding 2 ml of an aqueous solution at 0°C containing a 1000-fold excess of unlabeled amino acid corresponding to the [^{14}C]amino acid used, followed by 2 ml of cold aqueous 10% trichloroacetic acid (TCA). The labeled precipitate is then washed, heated to 90°C, and counted by conventional procedures.

Depending on the relative concentration of macromolecular components, S-60 extracts incorporate between 8 and 37 pmoles of [^{14}C]leucine per reaction mixture, corresponding in each case to about 140 pmoles of leucine per milligram of ribosomes. Comparable incorporations can be obtained for alanine, arginine, and valine; other amino acids give somewhat smaller incorporations except asparagine, histidine, and glutamine which incorporate poorly.

V. PREPARATION OF RIBOSOMES FREE OF mRNA AND OF S-150 EXTRACTS

Ribosomes are freed of endogenous mRNA by incubating S-60 extracts in bulk in a modified reaction mixture shown in Table 5. This mixture omits added amino acids and increases the relative proportion of S-60 extract to conserve other chemicals. After incubation the reaction mixture is cooled to 0°C and allowed to

TABLE 5

Reaction Mixture for the Bulk Incubation
of S-60 Extracts in the Preparation of
Preincubated Ribosomes

Component	Amount
0.03 M Sodium ATP	0.25 ml
0.12 M Sodium PEP	0.25 ml
0.012 M Sodium or lithium GTP	0.375 ml
1 M Tris-HCl buffer (pH 8.05)	0.125 ml
5 M NH_4Cl + 0.6 M $(NH_4)_2SO_4$	0.75 ml
Solid KCl	0.56 gm
S-60 Extract in solution D	2.00 ml
Total nominal volume[a]	3.750 ml

[a]The nominal final salt concentrations are 3.8 M KCl, 1.2 M NH_4^+, and 0.05 M magnesium acetate. In the development of the cell-free halophile system, this combination of salts preceded that given in Table 4 but has been retained here as for some unknown reason it produces more active ribosomes.

stand so that excess salt crystallizes. The liquid is then decanted from the crystals, diluted with solution D as required for filling ultracentrifuge tubes, and centrifuged at 150,000 X g for 150 min. The ribosomal pellets obtained are resuspended

3. Protein Synthesis from Halophilic Bacteria

in fresh solution D using a glass-Teflon homogenizer and then centrifuged again at 150,000 X g for 120 min. The pellets are resuspended in fresh solution D as before and the suspension is clarified by centrifuging at 40,000 X g for 10 min. The final suspension of preincubated ribosomes should have a volume of one-sixth to one-eighth of the volume of the S-60 extract from which it was derived and should have an A_{260} of about 350, corresponding to about 22 mg ribosomes per milliliter. It is convenient to freeze and store it in vials under liquid nitrogen in 0.25-ml aliquots.

S-150 supernatant extracts are obtained by centrifuging S-60 extracts at 150,000 X g for 2 hr to pellet the ribosomes. The supernatant obtained from preparing preincubated ribosomes can be used except that it has been diluted to twice its original volume in the reaction mixture and requires further dialysis to remove the low-molecular-weight additions.

VI. INCORPORATION WITH SYNTHETIC mRNAs

When incorporation is to be directed by synthetic mRNAs, the reaction mixture in Table 4 is used with the 20 µl of S-60 extract replaced by 15 µl of S-150 extract and 5 µl of pre-incubated ribosomal suspension. In addition, the polyribonucleotide in water replaces the 10 µl of water. An appropriate quantity of synthetic mRNA to use is 50 µg per reaction mixture, corresponding to 0.4 mg/mg of preincubated ribosomes. With this amount of poly U, the incorporation of [^{14}C]phenylalanine with specific activity of about 16-18 Ci/mole should be about 350 pmoles/mg of ribosomes. However, with [^{14}C]phenylalanine with specific activity of about 370 Ci/mole, incorporation is about half of this because of the much lower concentration of amino acid in the reaction mixture.

High-molecular-weight polyribonucleotides containing large percentages of adenine cannot be used as they are insoluble in the salt solutions.

VII. PREPARATION OF AMINOACYL-tRNA SYNTHETASES AND tRNA

An S-150 supernatant extract is prepared as described above except that the following changes can be made:

(1) Cultures can be grown for 36 hr to increase the yield of cells to between 4 and 6 g wet weight/liter of culture.

(2) Washed cells can be broken by either

(a) grinding them in a mortar for 5-10 min with an amount equal to their wet weight of acid-washed alumina (Alcoa alumina A-301). A volume of solution D corresponding to 1-1.5 times the wet weight of cells, together with β-mercaptoethanol and DNase at the same concentrations as before, is added and the mixture is ground for a few more minutes. This procedure is similar to that for E. coli.

or (b) suspending them in a volume of solution D plus β-mercaptoethanol corresponding to their wet weight and then

3. Protein Synthesis from Halophilic Bacteria

homogenizing in an ice-cold high-speed mixer, for example, a Servall Omnimixer at top speed for 30 sec. The homogenate is then incubated with DNase [1 mg/30 ml of solution D, added as a solution of 1 mg/ml in 0.1 M potassium acetate buffer (pH 5.0)] for 30 min at 37°C.

(3) The cell homogenate can be centrifuged at low speed to pellet unbroken cells and cell envelopes (and alumina if used) and then at 150,000 X g for 3 hr to sediment the ribosomes.

The above procedures also yield ribosomes, of course, which after washing by differential centrifugation can be used for physicochemical study.

The pH of the final S-150 supernatant extract is adjusted to 5 by slowly adding 1 M acetic acid. During this time and for about 20 min after pH 5 has been reached, the solution is stirred. The precipitate formed is sedimented, resuspended in solution D with β-mercaptoethanol as before, and dialyzed overnight against the same solution. After low-speed centrifugation to remove any remaining precipitate, the solution representing the crude synthetase preparation is frozen and stored in vials under liquid nitrogen.

Unlike the tRNA of nonhalophilic systems, that from H. cutirubrum does not precipitate with the synthetases at pH 5. To prepare tRNA from the pH 5 supernatant, the pH is first readjusted to 7.6 with 1 M KOH. The salt concentration is then reduced by dialyzing overnight against a large volume of

0.005 M tris-HCl buffer (pH 7.6) containing 1 µl of β-mercaptoethanol per 2 ml solution. Care must be taken to allow for a large increase in volume within the dialysis sac. After dialysis 0.01 vol of sodium dodecyl sulfate (0.125 gm/ml) is added and the mixture stirred for 20 min. tRNA is then isolated by the usual phenol extraction procedure (see Chapter 10) and precipitated by acid potassium acetate (pH 5, 20% w/v) and ethanol [16-18]. The precipitated tRNA is redissolved in 0.005 M KCl-0.005 M magnesium acetate-0.005 M tris-HCl buffer (pH 7.06) and reprecipitated with potassium acetate and ethanol as before. The final precipitate is washed with ethanol-water (2:1), then successively with ether-ethanol (1:2, 1:1, and 2:1), and finally with anhydrous ether, dried under vacuum, and stored as a dry powder at -20°C.

Alternatively, the tRNA can be freeze-dried as follows. The combined aqueous phases from phenol extraction are made 0.1 M with respect to NaCl by adding an appropriate volume of 2 M NaCl; 2 vol of ice-cold absolute ethanol are added. After stirring, the mixture is kept at -20°C in the deep-freeze compartment of a refrigerator overnight. The precipitate that forms is sedimented by centrifugation and redissolved in a suitable volume of 1 M potassium acetate (pH 5); 2 vol of cold ethanol are added and the mixture is left overnight at -20°C. This latter precipitation is repeated. The final precipitate is dissolved in 0.01 M KCl-0.01 M magnesium acetate-0.01 M

3. Protein Synthesis from Halophilic Bacteria

tris-HCl buffer and dialyzed exhaustively against glass-distilled water containing a small amount of β-mercaptoethanol (5 μl/20 ml). The tRNA solution is then freeze-dried.

VIII. MEASUREMENT OF AMINOACYL-tRNA FORMATION

The ability of an aminoacyl-tRNA synthetase preparation to catalyze the formation of aminoacyl-tRNA and of tRNA to accept amino acids is determined from the incorporation of [^{14}C]amino acid into cold-TCA-precipitable material. The reaction mixture contains the following components in 3.8 M KCl, 1.4 M NaCl, 0.04 M magnesium acetate, and 0.03 M tris-HCl buffer (pH 8): sodium ATP, 0.3 μmole; [^{14}C]amino acid (0.5-1 μCi); aminoacyl-tRNA synthetase preparation, 0.2-0.4 mg protein; tRNA, 0.1-0.3 mg. As for the reaction mixtures described above, dry KCl must be added so that the same qualifications as before apply to the calculated salt concentrations. The reaction mixture is given in detail in Table 6.

The reaction mixture is incubated at 37°C for 15-20 min. The reaction is terminated by adding in order: 2 ml of an aqueous solution at 0°C containing a 1000-fold excess of unlabeled amino acid corresponding to the [^{14}C]amino acid used; 0.1 ml of a solution of serum albumin (5 mg/ml) to act as carrier; 2 ml of cold aqueous 10% TCA. The precipitate is washed without heating and prepared for counting by standard procedures.

TABLE 6

Reaction Mixture for the Formation of Aminoacyl-tRNA

Component	Amount
0.03 M Sodium ATP	10 µl
0.29 M Tris-HCl buffer (pH 8)	20 µl
Solid KCl	41 mg
5.6 M NaCl	55 µl
[^{14}C]Amino acid (neutralized)	10 µl
Cold amino acid mixture less labeled amino acid (each 8 X 10^{-4}M)[a]	10 µl
Water	30 µl
tRNA in solution D	70 µl
Aminoacyl-tRNA synthetases in solution D	20 µl
Total nominal volume[b]	225 µl

[a]If not required, these can be replaced by water.

[b]See footnote to Table 4, except that here the nominal final salt concentrations are 3.8 M KCl, 1.4 M NaCl, 0.04 M magnesium acetate, and 0.03 M tris.

Halophile aminoacyl-tRNA synthetases have different salt requirements for optimal activity, but the conditions given here are satisfactory for most of them. The incorporation of amino acids is between 0.1 and 1 nmole per mg of tRNA, except for

3. Protein Synthesis from Halophilic Bacteria   109

asparagine and glutamine which are incorporated poorly. Because of the different K_m values of the synthetases, the extent of incorporation of different amino acids often depends on the amount of labeled amino acid used.

REFERENCES

[1] H. Larsen, in The Bacteria: a Treatise on Structure and Function (I. C. Gunsalus and R. Y. Stanier, eds.), Vol. 14, Academic Press, New York, 1962, 297.

[2] H. Larsen, Advan. Microbial Physiol., 1, 97 (1967).

[3] D. J. Kushner, Advan. Appl. Microbiol., 10, 73 (1968).

[4] J. H. B. Christian and J. A. Waltho, Biochim. Biophys. Acta, 65, 508 (1962).

[5] S. T. Bayley and E. Griffiths, Biochemistry, 7, 2249 (1968).

[6] E. Griffiths and S. T. Bayley, Biochemistry, 8, 541 (1969).

[7] S. T. Bayley and E. Griffiths, Can. J. Biochem., 46, 937 (1968).

[8] S. T. Bayley, in Biochemical Adaptation (F. P. Conte, ed.), University of Chicago Press, Chicago, in press.

[9] N. E. Gibbons, in Methods in Microbiology (J. R. Norris and D. W. Ribbons, eds.), Vol. 3B, Academic Press, New York, 1969, 169.

[10] S. N. Sehgal and N. E. Gibbons, Can. J. Microbiol., 6, 165 (1960).

[11] H. Onishi, M. E. McCance, and N. E. Gibbons, Can. J. Microbiol., 11, 365 (1965).

[12] M. B. Gochnauer and D. J. Kushner, Can. J. Microbiol., 15, 1157 (1969).

[13] M. W. Nirenberg, in Methods in Enzymology, Vol. 6 (S. P. Colowick and N. O. Kaplan, eds.), Academic Press, New York, 1963, 17.

[14] O. H. Lowry, N. J. Rosebrough, A. L. Farr, and R. J. Randall, J. Biol. Chem., 193, 265 (1951).

[15] S. T. Bayley and D. J. Kushner, J. Mol. Biol., 9, 654 (1964).

[16] K. S. Kirby, Biochem. J., 64, 405 (1956).

[17] G. von Ehrenstein and F. Lipmann, Proc. Nat. Acad. Sci. USA, 47, 941 (1961).

[18] K. Moldave, in Methods in Enzymology, Vol. 6 (S. P. Colowick and N. O. Kaplan, eds.), Academic Press, New York, 1963, 757.

Chapter 4

DNA-DEPENDENT, RNA-DIRECTED PROTEIN SYNTHESIS

Masaki Hayashi

Department of Biology
University of California, San Diego
La Jolla, California

I. INTRODUCTION 111

II. MATERIALS. 114

   A. Buffers 114

   B. Cells 114

   C. DNA-Dependent RNA Polymerase. 114

III. PROCEDURE. 115

IV. ASSAY FOR RNA AND PROTEIN SYNTHESIS IN THE
COUPLED SYSTEM 116

V. APPLICATION OF THE SYSTEM. 117

   REFERENCES 118

I. INTRODUCTION

The DNA-dependent, RNA-directed protein-synthesizing system is referred to as a coupled system because genetic transcription is coupled to translation in a test tube.

The essential components of the coupled system are (1) DNA template, (2) DNA-dependent RNA polymerase, (3) ribosomes,

Copyright © 1971 by Marcel Dekker, Inc. No part of this work may be reproduced or utilized in any form or by any means, electronic or mechanical, including xerography, photocopying, microfilm, and recording, or by any information storage and retrieval system, without the written permission of the publisher.

(4) soluble components of cell extracts, (5) ribonucleotide triphosphates (ATP, GTP, CTP, and UTP), (6) 20 amino acids, (7) Mg^2, and (8) NH_4Cl or KCl. Usually, an ATP-generating system [such as phosphoenolpyruvate (PEP) and PEP kinase] is added to the coupled system. The factors required for the translation of mRNA (initiation, elongation, and termination factors) may be added to the coupled system depending upon the purpose of the experiments, although the preparation described in this chapter contains, to some extent, all of these factors.

DNA directs RNA synthesis by DNA-dependent RNA polymerase, and the synthesized RNA in turn is translated into protein by ribosomes and soluble components extracted from bacterial cells. Incorporation of nucleotide monophosphates into RNA and amino acids into polypeptides can be measured by adding radioactive nucleotide triphosphates (^{14}C, ^3H, or ^{32}P) and amino acids (^3H, ^{35}S, or ^{14}C) to the reaction mixtures. Cold trichloroacetic acid (TCA)-precipitable nucleotide counts represent RNA synthesis, and hot TCA-precipitable amino acid counts are the measure of polypeptide synthesis. When it is necessary to measure the synthesis of a specific protein in the coupled system, either some other assay, such as enzyme activity or antigenicity of the protein, or physicochemical separation of the protein must be used.

Since it is not the purpose of this chapter to review the coupled systems described in the literature, we briefly refer

4. DNA-Dependent, RNA-Directed Protein Synthesis

to some of them with emphasis on the biochemically active protein products.

Nisman [1] described a preparation from Escherichia coli that synthesizes induced β-galactosidase, β-galactoside transacetylase, and alkaline phosphatase. Zubay's group [2-7] also developed a coupled system that synthesized β-galactosidase. Their system is capable of mimicking the in vivo regulatory mechanisms that control the expression of the lactose operon. Using T-even phage DNA as a template, Gold and Schweiger [8-11] synthesized α- and β-glucosyl transferases, lysozyme, and dCMP-deaminase. When SP82 (a Bacillus subtilis phage) DNA was used, they also detected the phage-specific dCMP-deaminase. Young and Tissière [12] and Young [13] reported the synthesis of α-glucosyl transferase from the T4 genome. The ØX-174 replicating form DNA (RF DNA)-dependent coupled system has been developed in our laboratory [14-18]. Product proteins from this coupled system were compared with in vivo ØX-174-specific proteins by polyacrylamide gel electrophoresis. Most of the ØX-174-specific proteins found in vivo were synthesized in the coupled system. When T7 DNA was used instead of RF DNA with E. coli RNA polymerase, gene 1 product (T7-specific RNA polymerase [19]) was synthesized [20]. The newly synthesized T7 RNA polymerase was biochemically active.

In the succeeding section, we describe the preparation developed in our laboratory.

II. MATERIALS

A. Buffers

TMK: 10^{-2} M tris-HCl, 10^{-2} M magnesium acetate, 6×10^{-2} M KCl, 6×10^{-3} M 2-mercaptoethanol; final pH 7.8

10 X TMA: 10^{-1} M tris-HCl, 10^{-1} M magnesium acetate, 6×10^{-1} M NH_4Cl, 6×10^{-2} M 2-mercaptoethanol; final pH 7.9

TMA 0.06: 10-fold dilution of 10 X TMA with H_2O

TMA 0.3: Same as TMA 0.06, but NH_4Cl concentration is 0.3 M

These buffers are kept at 4°C no longer than 1 month.

B. Cells

Escherichia coli Q13 is grown in antibiotics medium 3 (Difco), supplemented with 1% glucose and 0.01 M phosphate buffer (pH about 7.2), at 37°C with vigorous aeration. At midlog phase (A_{660} about 0.8 measured by a Zeiss spectrophotometer), the culture is quickly chilled to about 10°C, then centrifuged. Cells are washed once with TMK buffer by centrifugation; 1.0-1.2 gm of wet cells from a 1-liter culture are usually obtained. Cells are kept at -70°C without loss of activity for at least 6 months. Other strains of E. coli such as D10 or C should give active preparations.

C. DNA-Dependent RNA Polymerase

The enzyme used in the coupled system is purified from E. coli C BTCC 122 according to the procedure of Berg et al. [21].

4. DNA-Dependent, RNA-Directed Protein Synthesis

The amount of endogenous RNA polymerase varies from preparation to preparation, but most of the preparations require added RNA polymerase.

III. PROCEDURE

All operations are performed at 0-4°C. In a French press (at 12,000 psi) with 8 ml of TMK buffer, 8 gm of cells are disrupted. The resulting suspension is centrifuged 20 min at 20,000 X g; the supernatant is centrifuged 30 min at 30,000 X g. The following components are added to the top three-fourths of the 30,000 X g supernatant: 4×10^{-2} M tris-HCl (pH 7.8), 1.2×10^{-2} M PEP, 4×10^{-5} M each of 20 amino acids, 9×10^{-3} M 2-mercaptoethanol, 2.5×10^{-3} M ATP, 5×10^{-4} M GTP, and 50 µg/ml PEP kinase (Calbiochem, 251 units/mg). The mixture is incubated 80-90 min at 33°C, to eliminate endogenous mRNA activity, and dialyzed 10-14 hr against three changes of TMA 0.06 buffer. The dialyzed solution is centrifuged 10 min at 30,000 X g and ribosomes are precipitated from the supernatant by centrifugation for 2 hr at 105,000 X g in the no. 40 rotor of a Spinco L-20 preparative ultracentrifuge. The high-speed supernatant is centrifuged again for 10 hr at 105,000 X g. At the end of the run, the top 3 ml are discarded, and the next 9 ml are stored in small aliquots at -70°C as the soluble fraction (A_{260} nm about 25/ml).

The surface of the ribosome pellet (derived from a 2-hr spin at 105,000 X g) is washed with 1-2 ml of TMA 0.3 and the ribosome pellet is suspended in the same buffer (10 ml) and stored at 0-4°C overnight. The suspension is centrifuged 10 min at 30,000 X g, the pellets are discarded, and the ribosomes are precipitated from the supernatant by centrifugation for 90 min at 105,000 X g. The pellet is washed with TMA 0.06 buffer, resuspended in TMA 0.06, and centrifuged 10 min at 30,000 X g as before. The supernatant (ribosome fraction, A_{260} nm about 500/ml) is stored in small aliquots at -70°C. The soluble and ribosome fractions are stable at -70°C for at least 3 months.

IV. ASSAY FOR RNA AND PROTEIN SYNTHESIS IN THE COUPLED SYSTEM

Standard reaction mixture (0.13 ml) contains: tris-HCl (pH 7.8), 1.30 µmole; NH_4Cl, 9.0 µmole; magnesium acetate, 1.5 µmole; 2-mercaptoethanol, 0.9 µmole; PEP, 0.6 µmole; ATP, 0.25 µmole; GTP, CTP, and UTP (either CTP or UTP is labeled with ^3H, ^{14}C, or ^{32}P), 0.05 µmole each; 20 amino acids (one of the amino acids is labeled with ^3H, ^{35}S, or ^{14}C; a different isotope is used than the one for CTP or UTP), 4×10^{-3} µmoles each; pyruvate kinase, 5 µg; RNA polymerase, 1-5 µg; ribosomes, 5 A_{260} units; soluble fraction, 0.75 A_{260} units; DNA, 2-10 µg.

After incubation at 33°C, the reaction mixture is diluted to 2 ml with cold H_2O; 1.0 ml of the mixture is brought to a final TCA concentration of 5%, and the precipitate is collected

4. DNA-Dependent, RNA-Directed Protein Synthesis

on a glass-fiber filter (Whatman GF/C), and washed with 20 ml of ice-cold 5% TCA. After the filters are dried, they are counted in a liquid scintillation counter. This fraction (cold TCA-precipitable counts) corresponds to the UTP (or CTP) incorporation into RNA.

The other half of the reaction mixture is brought to a final TCA concentration of 7% and incubated for 20 min at 90°C; 0.05 ml of 5 mg/ml bovine serum albumin is added and the mixture is chilled to 0°C. The precipitate is collected by centrifugation. The pellet is dissolved in 0.2 N NaOH and the solution is brought to a final TCA concentration of 5% at 0°C. The precipitate is plated and counted as before. This fraction (hot TCA-insoluble counts) corresponds to the incorporation of amino acids into protein. An example of the kinetics of RNA and protein syntheses using ØX-174 RF DNA is shown in Figure 1.

V. APPLICATION OF THE SYSTEM

The system described in the preceding section has been used in our laboratory in order to study ØX-174 RF DNA- [14-18], T7 DNA- [20], and SV40 DNA- [22] primed reactions. Characterization of the product proteins is described in the references. These DNAs are active templates in the system. T4 and T2 DNAs are also active templates. Commercial preparations of calf thymus and salmon sperm DNA are moderately active templates (about one-half to one-third of the protein synthesis of RF

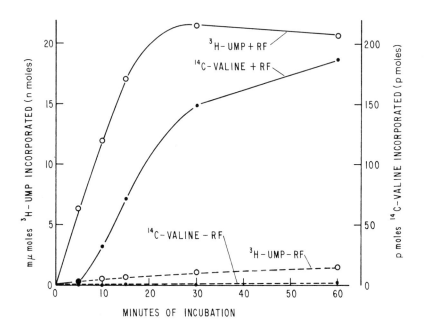

Fig. 1. Rate of RNA and Protein Synthesis with ØX-174 DNA.

or T7). Escherichia coli DNA prepared according to the method of Marmur [23], and λ DNA are poor templates. In addition to the DNAs described above, poly U and RNAs from the RNA bacteriophages (MS2, Qβ) are also active messenger RNAs in this system.

REFERENCES

[1] B. Nisman, in Methods in Enzymology, Vol. 12B (L. Grossman and K. Moldave, eds.), Academic Press, New York, 1968, p. 794.

[2] M. Lederman and G. Zubay, Biochim. Biophys. Acta, 149, 253 (1967).

4. DNA-Dependent, RNA-Directed Protein Synthesis

[3] J. K. DeVries and G. Zubay, Proc. Natl. Acad. Sci. U.S., 57, 1010 (1967).

[4] G. Zubay, M. Lederman, and J. K. DeVries, Proc. Natl. Acad. Sci. U.S., 58, 1669 (1969).

[5] G. Zubay and M. Lederman, Proc. Natl. Acad. Sci. U.S., 62, 550 (1969).

[6] M. Lederman and G. Zubay, Biochem. Biophys. Res. Commun., 32, 710 (1968).

[7] D. A. Chambers and G. Zubay, Proc. Natl. Acad. Sci. U.S., 63, 118 (1969).

[8] M. Schweiger and L. M. Gold, Cold Spring Harbor Symp. Quant. Biol., 34, 763 (1969).

[9] L. M. Gold and M. Schweiger, J. Biol. Chem., 244, 5100 (1969).

[10] L. M. Gold and M. Schweiger, Proc. Natl. Acad. Sci. U.S., 62, 892 (1969).

[11] M. Schweiger and L. M. Gold, Proc. Natl. Acad. Sci. U.S., 63, 1351 (1969).

[12] E. T. Young and A. Tissière, Cold Spring Harbor Symp. Quant. Biol., 34, 766 (1969).

[13] E. T. Young, J. Mol. Biol., 51, 591 (1970).

[14] R. N. Bryan, M. Sugiura, and M. Hayashi, Proc. Natl. Acad. Sci. U.S., 62, 483 (1969).

[15] D. H. Gelfand and M. Hayashi, Proc. Natl. Acad. Sci. U.S., 63, 135 (1969).

[16] R. N. Bryan and M. Hayashi, Biochemistry, 9, 1904 (1970).

[17] D. H. Gelfand and M. Hayashi, Proc. Natl. Acad. Sci. U.S., 67, 13 (1970).

[18] M. Hayashi, M. N. Hayashi, and H. Hayashi, Cold Spring Harbor Symp. Quant. Biol., 35, 174 (1970).

[19] M. Chamberlin, J. McGrath, and L. Waskell, Nature, 228, 227 (1970).

[20] D. H. Gelfand and M. Hayashi, Nature, 228, in press.

[21] P. Berg, K. Barrett, and M. Chamberlin, in Methods in Enzymology, Vol. 13 (J. M. Lowenstein, ed.), Academic Press, New York, 1968.

[22] R. N. Bryan, D. H. Gelfand, and M. Hayashi, Nature, 224, 1019 (1970).

[23] J. Marmur, J. Mol. Biol., 3, 208 (1961).

Chapter 5

RIBOSOMAL SUBUNIT EXCHANGE AND DENSITY GRADIENT CENTRIFUGATION

Raymond Kaempfer

The Biological Laboratories
Harvard University
Cambridge, Massachusetts

I. INTRODUCTION 121

II. PRINCIPLE OF THE TECHNIQUE 122

III. PROCEDURE. 124

    A. Isotopically Heavy and Light Polysome Extracts . 124

    B. Subunit Exchange Reaction. 126

    C. Role of Polysome Concentration 127

    D. Ultracentrifugal Analysis. 128

IV. APPLICATIONS 144

    A. Ribosomal Subunit Exchange 144

    B. Relative Rates of Subunit Exchange 144

    C. Other Applications 145

V. OTHER METHODS. 146

    REFERENCES 147

I. INTRODUCTION

During protein synthesis the two subunits of a functioning ribosome do not remain permanently associated. Instead, upon completing the translation of one or more polypeptide chains encoded by an mRNA molecule, ribosomes dissociate into their subunits. These subunits enter a pool of free ribosomal subunits. At the initiation of protein synthesis, ribosomal subunits are drawn from this pool and are recombined to form new ribosomes. Hence, between termination of one round of protein synthesis and initiation of the next, ribosomes undergo subunit exchange.

Here we describe the technique used in vitro to demonstrate that ribosomal subunit exchange occurs during protein synthesis and constitutes an essential part of its mechanism. This technique makes it possible to follow the movement of ribosomes through a single cycle of translation and therefore may be useful for studying various aspects of the mechanism of protein synthesis.

The demonstration and analysis of ribosomal subunit exchange in normally growing bacteria and yeast have been described elsewhere [1-3].

II. PRINCIPLE OF THE TECHNIQUE

Bacteria are uniformly labeled with heavy isotopes by growth in a medium containing ^{13}C, ^{15}N, and ^{2}H. An extract of these cells containing polysomes is mixed with a large excess of polysome extract from bacteria grown in normal (light) medium, and cell-free protein synthesis is allowed to occur. At intervals samples are digested with RNase to reduce polysomes

5. Ribosomal Subunit Exchange

to monosomes, and the density of the monosomes is analyzed by various ultracentrifugal methods. This procedure is outlined in Figure 1. Because of the predominance of light ribosomes in the reaction mixture, any heavy subunits generated by dissociation of the originally heavy ribosomes recombine almost exclusively with light subunits to form ribosomes of hybrid density. Thus subunit exchange in ribosomes is manifested by a progressive replacement of fully heavy ribosomes by two species of hybrid density: one containing a heavy 50 S and a light 30 S subunit, the other containing a light 50 S and a heavy 30 S subunit.

During the first round of subunit exchange, all heavy ribosomes are converted to hybrid species. Thereafter, the density distribution remains unchanged even though several more rounds of subunit exchange may occur.

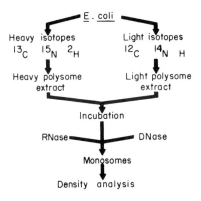

Fig. 1. Outline of procedure. (From Ref. 13.)

III. PROCEDURE

A. Isotopically Heavy and Light Polysome Extracts

A culture of Escherichia coli is grown at 37°C in 2-10 ml of light-isotope medium consisting of 0.5% Bacto-tryptone (Difco) and 0.25% NaCl, with [^{3}H]uracil (30 µCi/ml) or [^{32}P]PO$_4$ (15 µCi/ml) added to label ribosomes.

To grow cells uniformly labeled with heavy isotopes, a log-phase culture of E. coli growing in light medium is diluted with sterile H$_2$O and inoculated to a titer of approximately 10^5 cells per milliliter in 0.5 or 1 ml of heavy-isotope medium, consisting of D$_2$O (99.9% enriched, Bio-Rad) with 0.08 M tris-[^{2}H]HCl ([^{2}H]HCl should be used to adjust the pH to 7.4), 0.01 M NaCl, 0.005 M MgSO$_4$, 6 X 10^{-6} M FeCl$_3$, 10^{-6} M CaCl$_2$, 0.01 M [^{15}N]NH$_4$Cl (97.5% enriched, Miles), and sufficient [^{13}C, ^{15}N, ^{2}H] algal hydrolyzate (isotopic enrichments, 93%, 97.5%, and 99.9%, respectively) to support the growth of 5 X 10^8 cells per milliliter. Preparation of the algal hydrolyzate is described by Meselson and Weigle [4]. [^{32}P]PO$_4$ or [^{3}H]uracil is included to label ribosomes. In this medium the doubling time of the cells closely resembles that in light medium (40-50 min). Growth is followed by cell counting in a Petroff-Hausser chamber.

The density of the isotopically light cell culture should be adjusted so that both heavy and light cultures reach a titer

5. Ribosomal Subunit Exchange

of approximately 10^8 cells per milliliter at the same time. At this point each culture is rapidly mixed with 0.2 vol of ice in a tube previously held at -20°C. The cells are collected by centrifugation for 4 min at 5,000 X g and 2-4°C and resuspended in 0.13 ml of 25% sucrose in 0.01 M tris-HCl (pH 7.4). Polysome extracts are prepared immediately by neutral detergent lysis in the presence of 0.64% polyoxyethylene dodecyl ether (Brij 35, Atlas). This is accomplished by adding to the cell suspension 75 μl of lysozyme-EDTA mixture [composed of equal volumes of 0.85 mg/ml lysozyme solution in 0.01 M tris-HCl (pH 7.4) and 2.7 mg/ml EDTA solution, added together shortly before use]. After 60-70 sec the cell suspension is rapidly mixed with 0.23 ml of lysis buffer, which is made by combining 0.4 ml of H_2O, 0.1 ml of 5% Brij 35 in 0.01 M tris-HCl (pH 7.4), and 10 μl 1 M magnesium acetate. More than 90% of the cells are lysed by this procedure, as judged by loss of refractility under the microscope. To enhance the ribosome yield, sodium deoxycholate may be included in the lysis buffer or later added to the lysate, to a final concentration of 0.05% or less. All operations should be carried out at 0°C. Lysates should be kept at 0°C and used within 15 min for subunit exchange experiments. Alternatively, polysome lysates may be stored at -20°C and thawed once without significantly affecting ribosomal subunit exchange. This procedure is a modification of the method of Godson and Sinsheimer [5], which employs polyoxyethylene cetyl ether (Brij 58) as non-ionic detergent instead of the lauryl ether (Brij 35) used here. Brij 35 acts as efficiently as Brij 58 in bringing about cell lysis, with the advantage that 5% Brij 35 remains in solution at 0°C, whereas 5% Brij 58 readily precipitates at this temperature.

B. Subunit Exchange Reaction

The reaction mixture for subunit exchange experiments consists of 6.7 mM magnesium acetate, 60 mM NH_4Cl, 0.1 mg/ml of commercial E. coli tRNA, 0.2 mM ATP, 0.2 mM GTP, 5 mM phosphoenolpyruvate, 20 µg/ml of pyruvate kinase, 40 µM for each of 19 amino acids, 5 mM β-mercaptoethanol, 50 mM tris-HCl (pH 7.8); the final volume of 0.25 ml contains up to 50 µl of each polysome extract to be tested. The amounts of polysome extracts should be adjusted to give approximately a 1000-fold excess of light ribosomes over heavy ones. The mixture is brought to 37°C prior to the addition of polysome extracts; that from heavy cells is added last. The reaction is carried out at 37°C and terminated by adding either 8 µg of RNase, or sparsomycin (Cancer Chemotherapy National Service Center) to 10^{-4} M, and rapidly cooling the mixture to 0°C. (If sparsomycin is used, 8 µg of RNase should be added later.) Before density analysis, 8 µg of DNase is added.

This reaction mixture incorporates radioactive amino acids into trichloroacetic acid-precipitable material at a constant

5. Ribosomal Subunit Exchange

rate for at least 30 min. Amino acid incorporation is strongly inhibited in the presence of chloramphenicol or sparsomycin.

The reaction mixture is layered onto a preformed density gradient (Section III,D) which is placed in an International SB283 or a Spinco SW 40Ti rotor. After centrifugation fractions are collected on 1-1/4-in. Whatman no. 5 filter paper squares held in the orifice of scintillation vials. The filters are dried, pushed down into the vials, and counted by liquid scintillation spectrometry.

C. Role of Polysome Concentration

To obtain rapid and extensive ribosomal subunit exchange, it is necessary to use a concentrated polysome extract from light cells. The concentration of light polysomes in the reaction mixture should not be less than 10^8 cell equivalents per milliliter. This condition is met if the culture volumes and procedures described in Sections III,A and B are used. The effects of lower polysome concentrations are discussed in Section IV,C.

It is possible to observe ribosomal subunit exchange using fewer light polysomes if the reaction mixture is supplemented with S-30 from light cells. (S-30 is prepared from an RNase I-deficient strain, for example, D10 [6], by the method of Capecchi [7] and stored at -20°C.) In this case, however, the average size of the polysomes decreases rapidly during incubation. Thus, when polysome analysis is contemplated, the use of S-30 should be avoided.

D. Ultracentrifugal Analysis

That two reciprocal hybrid ribosome species result from ribosomal subunit exchange can be demonstrated by density equilibrium centrifugation in CsCl gradients [1]. In such gradients the two hybrid species are resolved in addition to fully heavy and fully light ribosomes. However, this technique is complicated by the fact that ribosomes are not stable in concentrated CsCl solutions. At high Mg^{2+} ion concentrations, ribosomes yield two bands in CsCl [8]. The denser band contains protein-deficient cores of 30 and 50 S subunits, while ribosomes are found only in the lighter band [9]. This phenomenon makes exact quantitation of hybrid formation difficult.

A more convenient approach is provided by the fact that heavy ribosomes and subunits sediment faster than their light counterparts in velocity gradients. The uniform substitution of ^{13}C, ^{15}N, and ^{2}H for ^{12}C, ^{14}N, and ^{1}H, respectively, imparts a mass increase of approximately 20% to ribosomal particles. This difference is sufficient to resolve fully heavy, hybrid, and fully light ribosomes by sedimentation velocity analysis, although individual hybrid species are only partly resolved. As shown in Section IV this technique allows reasonably accurate

5. Ribosomal Subunit Exchange

quantitation of the extent of ribosomal subunit exchange. Another important advantage of velocity sedimentation over density equilibrium centrifugation is that the former method allows the simultaneous separation of ribosomal particles according to mass and according to density.

1. Preparation of Preformed Density Gradients

Density gradients are necessary to obtain stably sedimenting bands during velocity sedimentation. We describe here procedures for making linear and exponential gradients and briefly discuss the properties of such gradients.

a. *Linear Gradients.* Preformed linear sucrose or glycerol gradients are most commonly used to stabilize sedimenting zones. Figure 2A schematically shows how such gradients are constructed by a slight modification of the method of Britten and Roberts [10]. With both stopcocks closed, vessel L is filled with half the final gradient volume of sucrose solution at the concentration desired at the top of the gradient. The connection between vessels H and L is briefly opened to expel air, and the excess sucrose in vessel H is returned by pipet to vessel L. Vessel H is then filled with half the gradient volume of sucrose solution at the concentration desired at the bottom of the gradient. The connection between the vessels is opened and, with stirring, the entire gradient volume is allowed to flow through a glass capillary or steel

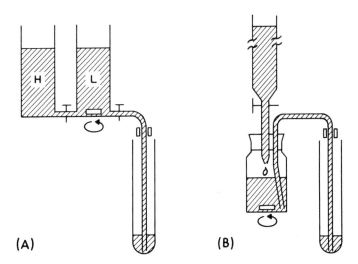

Fig. 2. (A) Apparatus for making linear gradients. (After Ref. 10.) (B) Apparatus for making exponential gradients. (After Ref. 11.) Their operation is described in the text.

needle into the bottom of a centrifuge tube. The gradient forms as the initial effluent is being displaced upward by an increasingly denser solution. Finally, the capillary is gently withdrawn from the centrifuge tube.

An alternative procedure is to reverse the concentrations of starting solutions in vessels H and L and to build the gradient by allowing the increasingly lighter effluent to flow down the wall of the centrifuge tube [10]. This procedure is not suitable for use with hydrophobic centrifuge tubes, such as those made of polypropylene or polyallomer, because the effluent flows down in the form of discrete drops which disturb the gradient.

5. Ribosomal Subunit Exchange

A disadvantage of linear gradients made with sucrose or glycerol is that the viscosity of their solutions increases considerably with concentration. This increase is so pronounced that the velocity of sedimenting particles actually decreases as they move down the gradient, even though the centrifugal acceleration to which they are subject increases. This deceleration diminishes the separation between two sedimenting zones.

b. <u>Exponential Gradients</u>. The problem of particle deceleration in linear sucrose (or glycerol) gradients can be largely overcome by the use of appropriately constructed exponential gradients. In these gradients the sucrose concentration increases more rapidly in the lighter half of the gradient than in the denser half, along a concentration curve which can be chosen so that the increase in viscous drag experienced by sedimenting particles is exactly compensated for by the increase in centrifugal acceleration acting on them. Hence in such gradients particles sediment with constant velocity at a rate strictly proportional to their standard sedimentation coefficient values (<u>isokinetic</u> gradients). The theory of constant velocity sedimentation and its validation have been described by Noll [11].

Furthermore, it is possible to construct exponential gradients in which particles sediment at a continually increasing velocity, leading to a separation greater than proportional to their standard sedimentation coefficient values [12]. These *accelerating* gradients are described in Section III,D,2,c.

Figure 2B schematically shows a simple device for making exponential gradients. A buret containing the heavier starting solution (sucrose concentration C_h) is connected through a rubber stopper to a mixing vessel containing a volume V_m of the lighter starting solution (sucrose concentration C_1). A needle passing through the rubber stopper into this solution is connected on the outside by a length of thin tubing to a glass capillary reaching to the bottom of a centrifuge tube. Sufficient manual pressure is exerted on the rubber stopper to fill the entire capillary with fluid, and the stopcock is opened. As the more concentrated solution drips into the mixing vessel, the gradient is formed upward from the bottom of the tube. During this process V_m remains constant. The solution in the mixing vessel is stirred magnetically. When the buret has delivered a volume equal to that of the desired gradient, the stopcock is closed and the capillary withdrawn. The mixing vessel is replaced by a fresh one containing V_m milliliters of sucrose at concentration C_1, and the process is repeated for the next gradient.

After V milliliters have dripped from the buret, the concentration C_v in the mixing vessel is [11]

$$C_v = C_h - (C_h - C_1) e^{-V/Vm} \qquad (1)$$

5. Ribosomal Subunit Exchange

Appropriate glassware for 12.5-ml gradients consists of a 50-ml buret with a Teflon stopcock, and standard glass scintillation vials as mixing vessels.

2. Resolving Power of Various Exponential Gradient Systems

a. *Sucrose-H_2O Gradients.* To make suitable exponential gradients for an International SB283 or Spinco SW 40Ti rotor, the following parameters may be used: $V = V_m = 12.5$ ml, $C_h = 25\%$ sucrose (w/v), $C_l = 12.5\%$ sucrose (w/v). Using the above procedure one obtains 12.5-19.6% exponential sucrose gradients as determined by refractive index measurements, values slightly less than the theoretical 12.5-20.4% calculated from Equation (1). These gradients are very nearly isokinetic.

Figure 3 depicts the sedimentation distribution of a mixture of heavy, hybrid, and light ribosomal particles after centrifugation in such a gradient. The gradient contains enzyme grade sucrose (Schwartz-Mann) in 0.01 M tris-HCl (pH 7.4), 0.01 M magnesium acetate, 0.05 M KCl, and 100 µg/ml of gelatin. The addition of gelatin serves to prevent adsorption of ribosomes to glass or to the walls of centrifuge tubes. It may be seen in Figure 3 that all species are well resolved: heavy ribosomes and subunits sediment approximately 23% faster than their light counterparts. This increase in sedimentation rate correlates well with the calculated mass increase imparted by the heavy isotopes.

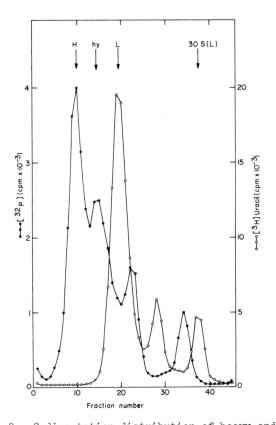

Fig. 3. Sedimentation distribution of heavy and light ribosomal particles in an exponential sucrose-H_2O gradient. ^{32}P-labeled heavy polysome extract was briefly incubated with excess ^3H-labeled light polysome extract under conditions described in Section III,B and sedimented through a 12.5-19.6% sucrose-H_2O gradient (12.5 ml). Centrifugation was for 4.1 hr at 41,000 rpm and 5°C in an SB283 International rotor. The total gradient contained 51 fractions. The peaks from right to left are: light 30 S [30 S (L)], heavy 30 S, light 50 S, and heavy 50 S subunits, light ribosomes (L), hybrid (hy) ribosomes, and heavy (H) ribosomes.

5. Ribosomal Subunit Exchange

In this sample partial ribosomal subunit exchange has occurred, as evidenced by the appearance of a hybrid (hy) ribosome peak. Since the 50 S subunit possesses approximately twice the mass of the 30 S subunit, subunit exchange in ribosomes should result in the appearance of two hybrid peaks, equally spaced between the positions of fully heavy and fully light ribosomes. The faster-sedimenting peak should contain heavy 50 S and light 30 S subunits; the slower peak is expected to contain heavy 30 S and light 50 S subunits. However, Figure 3 shows only one hybrid peak, sedimenting slightly slower than expected for the heavier hybrid, while the presence of the lighter hybrid is manifested only by a shoulder sedimenting between the main hybrid peak and light ribosomes. This pattern is explained by the fact that, with respect to the radioactive isotope originally present in heavy ribosomes, the heavier hybrid species contains approximately twice the amount of radioactivity as does the lighter hybrid, a ratio equal to the relative abundance of the labeled moiety, ribosomal RNA, in 50 and 30 S subunits.

Because light ribosomal particles are in a 1000-fold excess over heavy ones (Section III,B), radioactivity originally present in light ribosomes is not detectable in the hybrid region (see also Figure 4).

This gradient system has been used in studies of ribosomal subunit exchange in vivo [1,2] where it has also been shown

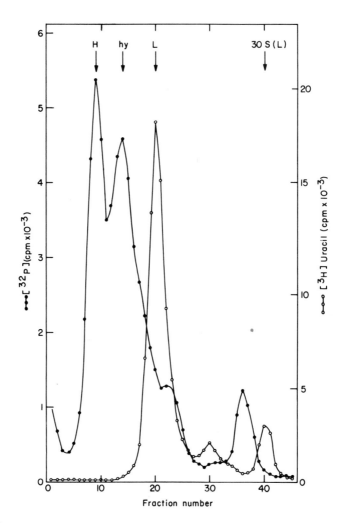

Fig. 4. Sedimentation distribution of heavy and light ribosomal particles in an exponential sucrose-D_2O gradient. Procedure as in Figure 3, except that D_2O (90% enriched) was substituted for H_2O in the gradient, and that centrifugation was for 6.0 hr. The total gradient contained 55 fractions. Peaks are as indicated in legend for Figure 3.

5. Ribosomal Subunit Exchange

that subunits constituting ribosomes of hybrid density have themselves remained fully heavy and fully light, respectively.

b. *Sucrose-D_2O Gradients.* Since the sedimentation coefficient of a particle is proportional to $(1 - \bar{v}\rho)/\eta$, where $\bar{v}$ is the partial specific volume of the particle in a solvent of density ρ and viscosity η, a rise in the density of the gradient solution retards particles with a lower density more than those with a higher density. Accordingly, the separation between heavy and light ribosomes can be improved by substituting D_2O for H_2O.

Although they are more viscous, exponential sucrose-D_2O gradients allow heavy ribosomal particles to sediment approximately 30% faster than the corresponding light ones (Figure 4). The lighter hybrid ribosome species is clearly indicated by the presence of ^{32}P counts between the main hybrid peak (fraction 14) and the heavy 50 S subunit peak (fractions 22 to 23). The isokinetic character of this gradient and the previous one (Figure 3) is illustrated by the positions of light and heavy 50 S subunit peaks, which are exactly equidistant to the 30 and 70 S peaks of corresponding density.

c. *$CsCl-D_2O$ Gradients.* Considerably improved separation between sedimenting particles can be obtained in gradients described below, in which the sedimentation velocity of a particle increases as a function of distance sedimented. In these gradients the faster-sedimenting particle is accelerated over the slower-sedimenting one, leading to a separation greater than proportional to their standard sedimentation coefficient values.

In these accelerating gradients, the viscosity <u>decreases</u> with increasing concentration. This is achieved by taking advantage of an unusual property of solutions of CsCl in H_2O or D_2O; at temperatures near 0°C, the viscosity of such solutions diminishes markedly with increasing CsCl concentration, up to a density of 1.5-1.6 g/cm^3. Consequently, in H_2O or D_2O stabilized by an appropriately chosen CsCl gradient, sedimenting particles are accelerated both by the increasing centrifugal force acting on them and by the decreasing viscous drag. The theory of sedimentation velocity analysis in accelerating gradients and the properties of such gradients are described elsewhere [3].

One type of accelerating gradient, in which sedimentation velocity increases almost 2.5-fold from top to bottom, is illustrated in Figure 5. This accelerating gradient is prepared according to the procedure described in Section III,D, 1,b, with $V = V_m = 12.5$ ml and starting solutions of concentrations $C_l = 0\%$ and $C_h = 20\%$ (w/w) CsCl (biological grade) in D_2O (90% enriched, Bio-Rad) containing 0.01 M magnesium acetate, 0.01 M tris-HCl (pH 7.4), and 100 μg/ml of gelatin. The resulting exponential gradient is 0-12.6% CsCl and has a density range from 1.095 to 1.223 g/cm^3. Alternatively, 5 and 25% (w/w)

5. Ribosomal Subunit Exchange

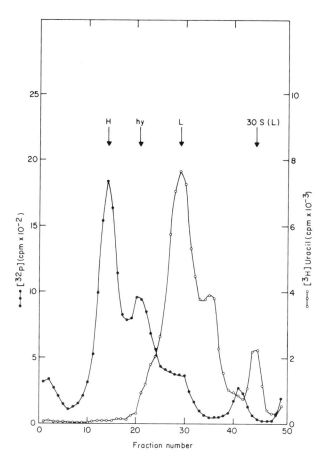

Fig. 5. Sedimentation distribution of heavy and light ribosomal particles in an exponential CsCl-D_2O gradient. Incubation procedure as in Figure 3, except for light polysome concentration (see text). The gradient, 0-12.6% CsCl in D_2O (90% enriched), was centrifuged for 2.5 hr at 41,000 rpm and 5°C in an SB283 International rotor. The total gradient contained 60 fractions. Peaks are as indicated in legend for Figure 3.

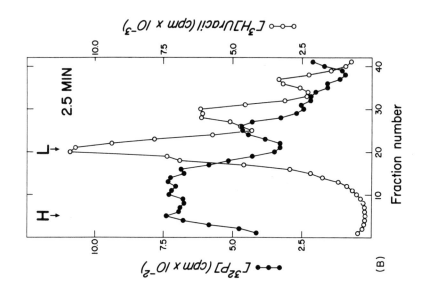

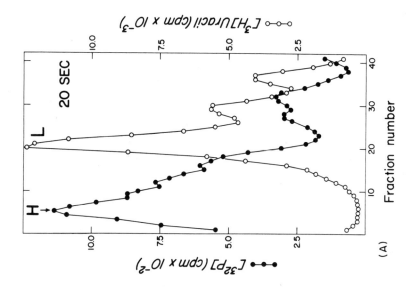

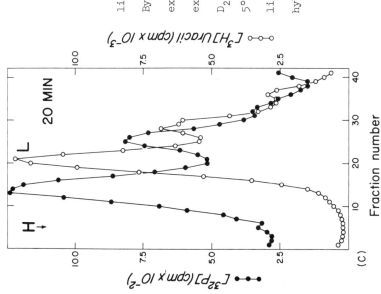

Fig. 6. (A-C). Sedimentation distribution of heavy and light ribosomal particles after cell-free protein synthesis. By means of the plan of Figure 1, ^{32}P-labeled heavy polysome extract was incubated with excess ^{3}H-labeled light polysome extract for the indicated time and examined on a 5-17.6% CsCl-D_2O gradient. Centrifugation was for 2.7 hr at 41,000 rpm and 5°C. The peaks from right to left are: light 30 S, heavy 30 S, light 50 S, and heavy 50 S subunits, light (L) ribosomes, hybrid ribosomes, and heavy (H) ribosomes. (From Ref. 13.)

CsCl starting solutions may be used to generate 5-17.6% exponential CsCl-D_2O gradients (density range, 1.146-1.272 g/cm^3). Ribosomes are quite stable in these gradients. D_2O remaining in the mixing vessel may be recovered by distillation.

Comparison with isokinetic sucrose gradients (Figures 3 and 4) shows that the main effect of this gradient is to greatly increase the separation between heavy and light ribosomes, while reducing the separation in the subunit region. This is easily appreciated by comparing in each gradient the distance between fully heavy (H) and fully light (L) ribosome peaks with the distance between the fully light ribosome peak and the light 30 S subunit peak.

Whereas a 1000-fold excess of light polysomes over heavy ones was used in the samples displayed in Figures 3 and 4, the excess of light polysomes used in the sample of Figure 5 was 15 to 20 times lower. Accordingly, the two shoulders of ^3H label seen in the hybrid region may reflect the presence of the two hybrid species, the lighter hybrid being twice as radioactive as the heavier one with respect to radioactivity originally present in fully light ribosomes.

Further examples of accelerating gradients are shown in Figure 6A, B, and C and Figure 7A, B, and C. The bandwidths of the peaks in these gradients are not much greater than those in isokinetic sucrose gradients. Therefore the increase in separation represents an actual increase in resolution. Note

5. Ribosomal Subunit Exchange

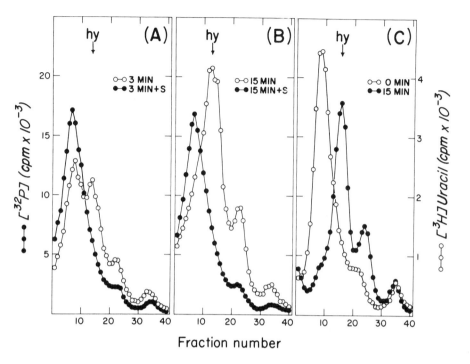

Fig. 7. (A-C). Effect of sparsomycin on ribosomal subunit exchange. ^{3}H-labeled heavy polysome extract was incubated with excess unlabeled light polysome extract for 0, 3, and 15 min. ^{32}P-Labeled heavy polysome extract was incubated with excess unlabeled light polysome extract for 15 min, and in the presence of 10^{-4} M sparsomycin (S) for 3 min and 15 min. The ^{3}H- and ^{32}P-incubation mixtures were then pooled as indicated in the figure and sedimented through 0-12.6% CsCl-D_2O gradients. Centrifugation was for 2.7 hr at 41,000 rpm and 5°C. Arrow indicates expected hybrid position. (From Ref. 13.)

that the times of centrifugation needed for $CsCl-D_2O$ gradients (Figures 5-7) are shorter than those needed for sucrose gradients (Figures 3 and 4).

IV. APPLICATIONS

A. Ribosomal Subunit Exchange

A ^{32}P-labeled polysome extract from cells grown in heavy-isotope medium is incubated with a 1000-fold excess of ^{3}H-labeled light polysome extract in the complete reaction mixture described in Section III,B. After 20 sec of incubation, the reaction is terminated and the density of the ribosomes is analyzed on a 5-17.6% exponential $CsCl-D_2O$ gradient (Section III,D,2,c). As can be seen in Figure 6A, during this time part of the ^{32}P label, which originally was present in a single, symmetrical fully heavy ribosome peak, has been displaced to the region between heavy and light ribosomes, where hybrid ribosomes are expected. The hybrid peaks increase rapidly with time, and at 2.5 min (Figure 6B) approximately half of the originally heavy ribosomes have been converted to hybrids. By 20 min (Figure 6C) few, if any, heavy ribosomes remain. Rapid and extensive subunit exchange has occurred.

B. Relative Rates of Subunit Exchange

A sensitive method for measuring relative rates of ribosomal subunit exchange is to use polysome extracts from

5. Ribosomal Subunit Exchange

two cultures grown in heavy-isotope medium, one labeled with [^{32}P]PO$_4$, the other with [^{3}H]uracil. Polysome extracts of these cultures are incubated in two separate tubes with excess nonradioactive extract from light cells. After termination of the reaction, the contents of the two tubes are pooled and sedimented through one gradient.

Figure 7 illustrates this approach. As seen in Figure 7A, after 3 min of incubation in the presence of sparsomycin, a potent inhibitor of peptide bond formation, none of the ^{32}P-labeled heavy ribosomes have undergone subunit exchange, although by this time substantial hybrid formation has occurred in the ^{3}H control. Even after 15 min a time sufficient for almost complete subunit exchange in the ^{3}H control, there is no evidence that any hybrid ribosomes are formed in the presence of sparsomycin (Figure 7B). The control (Figure 7C) shows that ^{32}P-labeled ribosomes are fully capable of undergoing subunit exchange. Thus sparsomycin completely blocks exchange of ribosomal subunits, demonstrating the absolute dependence of this process on protein synthesis.

C. Other Applications

The hybrid assay may be employed to study the mode of action of inhibitors (for examples, see Kaempfer [13] and Kaempfer and Meselson [14]). The heavy-isotope technique can also be used to study the role of subunit exchange in the formation of single ribosomes that accumulate in vivo and in vitro under various suboptimal conditions for protein synthesis [15].

The distribution of ribosomal particles among polysomes, single ribosomes, and subunits depends critically on the concentration of light polysome extract in the subunit exchange reaction mixture. As pointed out in Section III,C, rapid and extensive ribosomal subunit exchange, such as that illustrated in Figure 6, occurs in the presence of a concentrated light polysome extract. When the concentration of polysomes is lowered, the efficient maintaintance of polysome size is impaired and, instead, single ribosomes accumulate at the expense of polysomes [15]. By further diluting the polysomes several orders of magnitude, it is possible to attain subunit concentrations that limit the rate of re-formation of ribosomes. Hence in these dilute conditions ribosomal subunits accumulate at the expense of polysomes in a process strictly dependent on protein synthesis. Both polysomes carrying endogenous mRNA [14,15] or phage RNA [16] may be converted to ribosomal subunits in this way.

V. OTHER METHODS

A modification of the procedure described here was used by Guthrie and Nomura [17]. It entails growing cells in heavy medium containing per liter: 6.3 g Na_2HPO_4, 2.7 g KH_2PO_4,

5. Ribosomal Subunit Exchange

0.2 g $^{15}NH_4Cl$, 10^{-6} M $FeCl_3$, 10^{-3} M $MgSO_4$, 0.2% deuterated algal hydrolyzate (Merck, Sharp and Dohme, Canada), and D_2O (85% enriched). While this approach allows the preparation of large amounts of heavy cells, the resolution of their ribosomes from hybrid or fully light ones is rather limited, both in sedimentation velocity analysis and in density equilibrium centrifugation.

Ribosomal subunit exchange can also be detected without the use of heavy isotopes. Instead of measuring the conversion of heavy ribosomes to hybrids that contain one heavy and one light subunit, the rate of equilibration of label between unlabeled light ribosomes and radioactively labeled light ribosomal subunits is examined. Since in the latter method newly formed ribosomes cannot be separated from the original ones, the actual fraction of ribosomes that have undergone subunit exchange cannot be measured, but the net extent of subunit exchange can be calculated from the specific radio-activities of various ribosomal particles. This approach has been used in vitro to detect exchange of ribosomal subunits with mammalian polysomes [18] and with mammalian single ribosomes [18,19].

REFERENCES

[1] R. Kaempfer, M. Meselson, and H. Raskas, J. Mol. Biol., 31, 277 (1968).

[2] R. Kaempfer, Nature, 222, 950 (1969).

[3] R. Kaempfer, in Methods in Enzymology, Vol. 20 (L. Grossman and K. Moldave, eds.), Academic Press, New York, 1971, Chap. 48.

[4] M. Meselson and J. Weigle, Proc. Natl. Acad. Sci. U.S., 47, 357 (1961).

[5] G. Godson and R. Sinsheimer, Biochim. Biophys. Acta, 149, 476 (1967).

[6] R. Gesteland, J. Mol. Biol., 17, 67 (1966).

[7] M. Capecchi, Proc. Natl. Acad. Sci. U.S., 55, 1517 (1966).

[8] S. Brenner, F. Jacob, and M. Meselson, Nature, 190, 576 (1961).

[9] M. Meselson, M. Nomura, S. Brenner, C. Davern, and D. Schlessinger, J. Mol. Biol., 9, 696 (1964).

[10] R. J. Britten and R. B. Roberts, Science, 131, 32 (1960).

[11] H. Noll, Nature, 215, 360 (1967).

[12] R. Kaempfer and M. Meselson, in Methods in Enzymology, Vol. 20 (L. Grossman and K. Moldave, eds.), Academic Press, New York, 1971, Chap. 56.

[13] R. Kaempfer, Proc. Natl. Acad. Sci. U.S., 61, 106 (1968).

[14] R. Kaempfer and M. Meselson, Cold Spring Harbor Symp. Quant. Biol., 34, 209 (1969).

[15] R. Kaempfer, Nature, 228, 534 (1970).

[16] M. Grubman and D. Nakada, Nature, 223, 1242 (1969).

5. Ribosomal Subunit Exchange

[17] C. Guthrie and M. Nomura, Nature, 219, 232 (1968).

[18] G. Howard, S. Adamson, and E. Herbert, J. Biol. Chem., 245, 6237 (1970).

[19] A. Falvey and T. Staehlin, J. Mol. Biol., 53, 21 (1970).

Chapter 6

CHAIN INITIATION FACTORS FROM Escherichia coli

Jerold A. Last*

The Rockefeller University
New York, New York

I. INTRODUCTION 151

II. SOURCES. 154

III. PREPARATION OF CRUDE FACTORS 154

    A. Purified f_1. 159

    B. Purified f_2. 163

    C. Purified f_3. 167

IV. ASSAYS 169

    A. Assays for f_1. 170

    B. Assays for f_2. 171

    C. Assays for f_3. 172

    REFERENCES 174

I. INTRODUCTION

Initiation factors from Escherichia coli were discovered independently by groups at New York University [1], Yale University [2], and the Institut de Biologie physico-chimique [3]. These factors are required for the translation of

*Present address: National Academy of Sciences, Washington, D.C.

Copyright © 1971 by Marcel Dekker, Inc. No part of this work may be reproduced or utilized in any form or by any means, electronic or mechanical, including xerography, photocopying, microfilm, and recording, or by any information storage and retrieval system, without the written permission of the publisher.

natural mRNA molecules when well-washed (with buffers that contain 0.5-2 M salt) ribosomes and physiological concentrations of Mg^{2+} ion are used in cell-free extracts that synthesize polypeptides. The factors promote the transfer of N-formylmethione residues from N-formylmethionyl-tRNA (fMet-tRNA) into the N-terminal position of synthesized polypeptides. They also increase the binding of fMet-tRNA to purified ribosomes in the presence of the trinucleotide AUG or of oligonucleotides containing the AUG triplet at or near their 5'-end. Under the same conditions the factors have no effect on (or slightly inhibit) the codon-directed binding of noninitiator aminoacyl-tRNA molecules to washed ribosomes. It is noteworthy that the factors are not required for the translation of some synthetic polynucleotides such as poly U and poly A, or when high concentrations of Mg^{2+} ion (14-18 mM) are present.

Initiation factors can be prepared from either the supernatant fraction after ribosomes are pelleted or, preferably, from the solutions remaining after crude ribosomal preparations are washed. Such wash fluids, or supernatant fractions, either before or after concentration by precipitation with $(NH_4)_2SO_4$, are referred to as crude preparations of initiation factors. There are at least three factors present in such a crude preparation, for which we use the nomenclature f_1, f_2, and f_3. Synonymous terms are as indicated in Table 1. The sometimes confusing terminology is attributable to the historical accident

6. Chain Initiation Factors from Escherichia coli

Table 1

Terminology of Initiation Factors

Factor	Reference
f_1	[7]
A	[8]
f_2	[7]
C	[8]
f_3	[7]
B	[8]
Brown and Doty factor (?)	[9]
Crude initiation factors	[2]

of independent discovery. In addition, f_1 and f_2 were named for their order of elution from a DEAE-cellulose column; f_3, which was discovered later, is actually eluted from such a column at a lower salt concentration than is f_2. Thus f_1, f_2, and f_3 correspond to A, C, and B, respectively. They are clearly different than (and have been separated from) chain elongation factors (Chapter 7) and chain termination factors (Chapter 8).

II. SOURCES

In addition to E. coli, factor preparations have been reported from Bacillus stearothermophilus [4], reticulocyte ribosomes [5], and ribosomes from embryonic chick muscle [6]. The detailed methods for their preparation, which are beyond the scope of this chapter, are available in the references cited (see also Vol. 2 of this series). All these preparations are modeled on the E. coli work; washing of the ribosomes with a buffer of high salt concentration, $(NH_4)_2SO_4$ precipitation, and (in some cases) chromatography on DEAE-cellulose and/or Sephadex columns are common features of the preparations.

III. PREPARATION OF CRUDE FACTORS

The usual source is the soluble fraction that results from washing ribosomes with high concentrations of NH_4^+ or K^+ (0.5-2.0 M) in buffer [1]. An alternative preparation starts with the lower one-third of the supernatant fraction after ribosomes are pelleted in a centrifuge. This fraction is rich in 30 S ribosomal subunits and DNA [3,10]. In earlier studies $(NH_4)_2SO_4$ precipitation was used only to concentrate the factors; a broad range of concentration (usually about 30-80% of saturation) of the salt was used. Recently, the differential solubility of f_2 and $f_1 + f_3$ in $(NH_4)_2SO_4$ solutions has been exploited to give an initial fractionation of these activities prior to chromatographic purification. A typical procedure

6. Chain Initiation Factors from Escherichia coli

for the preparation of crude initiation factors is as follows.

Bacteria are grown at 37°C to late-log phase in any of a number of different media [nutrient broth, trypticase-soy broth (both Difco products), or salts and glucose (supplemented with Difco casamino acids) have all been used successfully]. Escherichia coli strains K12, B, Q-13, and MRE 600 (the last-mentioned two mutants are deficient in certain RNase activities) are all sources of factors. The bacteria are harvested by centrifugation (all operations are carried out at 0-4°C from this point onward, unless otherwise noted). The cell paste is washed once by resuspension in 0.2 M NH_4Cl (pH 7.0), centrifuged (at about 10,000 rpm in a Servall small rotor), washed once with buffer A [20 mM tris-HCl (pH 7.8), 10 mM magnesium acetate, 10 mM 2-mercaptoethanol], and again collected by centrifugation. This is a convenient point in the preparation to store the cells in a freezer as active factors may be prepared from cells frozen for many months.

Washed cells are ground in a chilled mortar with twice their weight of alumina (Alcoa A-301 is convenient); 20- to 30-g batches of cells can be ground efficiently without over-loading the usual laboratory centrifuge. It is difficult to describe the correct end point for cell grinding; the alumina and cell mixture should assume a consistency of chewing gum and make distinct popping sounds. Addition of a few milliliters of buffer after 5-10 min is sometimes helpful, but too much should not be added or the paste will become too slippery for effective cell grinding. Cells should not be ground too long as degradation of proteins may result. The cell-alumina paste is extracted with 2 vol of buffer A (assume a density of 1.0). Cell fragments and alumina are removed by centrifugation at 30,000 X g for 30 min. (An alternative scheme for cell rupture by sonic disintegration has been described in Ref. 12.) The supernatant is treated with 3 μg/ml DNase (Worthington, electrophoretically purified; cruder preparations of DNase contain contaminating RNase and are obviously unsuitable for this purpose.) for a few minutes in the cold, then centrifuged for 30 min at 30,000 X g (Servall angle-head rotor), to give a preparation referred to as S-30. Ribosomes are pelleted from the S-30 fraction by centrifugation in a Spinco Model L ultracentrifuge for 4 hr at 93,000 X g (type 40 rotor) or 2.5 hr at 160,000 X g (type 50 rotor). The upper two-thirds of the supernatant, called S-100 or S-160, is aspirated off, dialyzed overnight against buffer A, and frozen in small aliquots as a source of transfer enzymes, aminoacyl-tRNA synthetases, and so on. The lower one-third of the supernatant is usually discarded. (However, preparation of factors from this lower one-third of the S-100 has been described [10], as well as preparation from the lower one-third of S-150 [3] or whole S-100 [11].) The ribosomal pellet is thoroughly drained.

6. Chain Initiation Factors from Escherichia coli

Initiation factors are now prepared by washing the drained ribosomes. In the earlier literature 0.5 M NH_4Cl in buffer was used to extract the factors. Present methods use 1.0 M or even 2 M NH_4Cl for this purpose, to ensure the extraction of f_3. The ribosomal pellets are suspended in buffer B [1.0 M NH_4Cl, 20 mM tris-HCl (pH 7.8), 2 mM magnesium acetate, 2 mM dithiothreitol (DTT)]; about 1 ml of buffer B for each 2 g of cells in the starting preparation is used. Alternatively, about 10 ml of buffer B are used for the amount of ribosomal pellet in each tube of a Spinco 40 rotor. The suspension is stirred (using a magnetic stirrer) overnight in a beaker in a cold room. The resulting suspension is centrifuged for 20 min at 30,000 X g to remove aggregated material, then at 93,000 X g for 4 hr or 160,000 X g for 2.5 hr. The washed ribosomes are suspended in buffer that contains 250 mM NH_4Cl, 5 mM magnesium acetate, 20 mM tris-HCl (pH 7.8), 0.5 mM DTT, and 50% (v/v) glycerol, at a concentration of about 1000 A_{260} units/ml. Under these conditions the ribosomes (which may be further purified by chromatography on DEAE-cellulose before storage [13]) can be stored for at least 1 month at -10°C (the glycerol prevents this solution from freezing) with no loss of activity.

The supernatant from the 1.0 M NH_4Cl wash of the ribosomes is fractionated by precipitation of proteins with $(NH_4)_2SO_4$. Finely powdered $(NH_4)_2SO_4$ is slowly added to a beaker (a magnetic stirrer should be used) containing an NH_4Cl wash solution (about 10-15 mg/ml of protein). The first fraction is taken when 19.4 g/100 ml (33% saturation) of $(NH_4)_2SO_4$ have been added. The solution is allowed to equilibrate for 20-30 min, then centrifuged for 10-15 min at 30,000 X g (Servall). The precipitate is discarded. Additional $(NH_4)_2SO_4$ (5.7 g/100 ml; to 42% saturation) is added slowly as above; the solution is equilibrated and centrifuged exactly as described. The precipitate is saved for the preparation of f_2. Additional $(NH_4)_2SO_4$ (22.1 g/100 ml; to 70% saturation) is added and the solution is treated exactly as described above. The precipitate is used as a source of f_1 and f_3. (There may be other initiation factors in this fraction [12], but such preparations are beyond the scope of this chapter.) The supernatant is discarded.

The procedure used in earlier studies, which does not separate f_2 from other factors, has the advantage of simplicity (especially of assay). Since f_1, f_2, and f_3 can be separated on a DEAE-cellulose column, the simple procedure is useful for all purposes except preparation of large batches of highly purified factors. The method described below is the source of crude factors for the DEAE-cellulose step described in Section III,A.

The supernatant solution from the 1.0 M NH_4Cl wash of the ribosomes is again used as the source of factors. $(NH_4)_2SO_4$

6. Chain Initiation Factors from Escherichia coli

(21 g/100 ml; 35% saturation) is added to the supernatant and the resulting solution is mixed and centrifuged as described above. The precipitate is discarded. Additional $(NH_4)_2SO_4$ (18.4 g/100 ml; 60% saturation) is added, the solution is mixed and centrifuged, and the supernatant is discarded. The precipitate, which contains the factors, is dissolved in buffer C [0.01 M KH_2PO_4-K_2HPO_4 (pH 7.5), 0.01 M 2-mercaptoethanol] at a concentration of about 10 mg/ml and dialyzed overnight against buffer C to remove residual $(NH_4)_2SO_4$.

A. Purified f_1

The crude preparation of initiation factors, concentrated by precipitation with $(NH_4)_2SO_4$, is further purified by passage through a cationic ion-exchange column. Usually, DEAE-cellulose is used; DEAE-Sephadex may be substituted at this step. The factor f_1 is a basic protein of low molecular weight with very little affinity for the positively charged groups on the column. Thus f_1 is eluted from the column with the washing buffer (containing 0.025 M NH_4Cl). The other factors, f_2 and f_3, require much higher concentrations of salt (0.10-0.25 M NH_4Cl) in the buffer to be eluted from DEAE-cellulose columns. The f_1-containing eluate is relatively highly purified; at this step the fraction is free of transfer enzymes (T_S, T_U, and G), nucleases, aminoacyl-tRNA synthetases, and f_2.

Approximately 100 mg of the protein from the 30-70% saturated $(NH_4)_2SO_4$ fraction, in 10 ml of 0.01 M KH_2PO_4-K_2HPO_4

(pH 7.5)-0.01 M 2-mercaptoethanol (buffer C), is applied at 0.4-0.5 ml/min to a column (1.1 X 36 cm) of DEAE-cellulose (0.93 meq/g or 0.88 meq/g Bio-Rad gave successful preparations) equilibrated with buffer C. (There is batch-to-batch variation in DEAE-cellulose. Wash well, remove fines, and test on a pilot scale before use. Some batches are better than others for yield and purity of f_2 and f_3.) The column is washed with 50 ml of 0.025 M NH_4Cl in buffer C to elute f_1. Further details of the development of this column are given in Section III,B. The f_1 in the eluate is concentrated by precipitation with $(NH_4)_2SO_4$ to 90% saturation, then dialyzed overnight against 0.01 M tris-HCl (pH 7.8)-0.01 M 2-mercaptoethanol (buffer D). This preparation of f_1 is stable for at least several months in a freezer (-10 to -20°C). In the presence of higher concentrations of salts, f_1 is rather heat stable; it loses 90% of its activity (MS2-directed lysine incorporation) in 5 min at 70°C in buffer D but loses only 10-20% of its activity in 5 min at 85°C in 0.08 M NH_4Cl-0.06 M tris-HCl (pH 7.8).

A variation on this procedure in which f_1 was purified by virtue of its lack of affinity for a column of DEAE-Sephadex has been reported [14]. Another minor variation, for convenience in scaling up the preparation, was the insertion of a liquid-liquid partition step to remove nucleic acids and some extraneous proteins before precipitation of crude factors with $(NH_4)_2SO_4$ [15].

6. Chain Initiation Factors from Escherichia coli

Initiation factor f_1, after the DEAE-cellulose step, can be further purified by absorption on a column of cellulose phosphate, also called phosphocellulose (Whatman P11). The active f_1 is eluted from such an anionic column by 0.5 M KH_2PO_4-K_2HPO_4 (pH 7.2) [8,16] by an NH_4Cl gradient (f_1 is eluted at about 0.4 M salt). The active fractions from phosphocellulose are pooled, concentrated, and passed through a column of Sephadex G-50 to give a preparation that is over 90% pure (only a single band is observed on gel electrophoresis).

The f_1 that has passed through DEAE-cellulose, been concentrated by precipitation with $(NH_4)_2SO_4$, and dialyzed is further purified on a column of phosphocellulose as follows: 82 mg of f_1 in 420 ml of buffer G [0.2 M tris-HCl (pH 7.4), 1 mM EDTA, 0.5 M DTT, 10% (w/v) glycerol (glycerol, and in the case of f_3, phosphate also, acts in some unknown way, probably related to the stabilization of subunits, to make some proteins more stable; hence, it is included in some buffers.)] is passed through a column (1.5 X 50 cm) of Whatman P11 (7.2 meq/g). The column is further washed with 40 ml of 0.25 M NH_4Cl in buffer G. A linear gradient from 0.25 to 0.75 M NH_4Cl in buffer G (500 ml) is used to elute f_1. A flow rate of about 20 ml/hr is used; fractions of about 3 ml are collected. The f_1 elutes from the column at about 0.4 M salt concentration. The active fractions are pooled, concentrated by precipitation with $(NH_4)_2SO_4$, and further purified by chromatography on Sephadex

G-50 at room temperature. The f_1 solution (5.5 mg in 1.5 ml of 0.20 M NH_4Cl in buffer G) is added to a column (95 X 0.95 cm; 82 ml) of Sephadex G-50. The column is eluted at a flow rate of 7 ml/hr with 0.20 M NH_4Cl in buffer G. Fractions of about 1.5 ml are collected. The f_1 is eluted from the Sephadex about 25 ml after the void volume and coincides with a sharp peak of absorbance at 280 nm. This f_1 fraction gives only a single band of protein on polyacrylamide gel electrophoresis at pH 4.3 and 8.7 [17].

An alternative purification scheme for f_1 after passage through DEAE-cellulose is as follows [18]. The pooled fractions are heated at 65°C for 5 min. Denatured protein is removed by centrifugation; the supernatant is passed through a column of carboxymethyl cellulose (a cation-exchanger cellulose). The f_1 is eluted from this column by a linear gradient from 0 to 0.35 M NH_4Cl in tris-HCl buffer (pH 7.4). The active fractions are pooled, the NH_4Cl concentration is raised to 0.2 M, and the solution is applied to a phosphocellulose column. The f_1 is eluted with a linear gradient from 0.2 to 0.7 M NH_4Cl in tris-HCl buffer (pH 7.4). This preparation of f_1 shows only one band on polyacrylamide gel electrophoresis at pH 4.5 in 8 M urea; it is purified 18.4 times with a 47% **recovery** of activity, compared to f_1 that has been passed through DEAE-cellulose.

Another variant of the purification scheme for f_1 after

6. Chain Initiation Factors from <u>Escherichia coli</u>

the DEAE-cellulose step is as follows [19]. The concentrated preparation is passed through a column of Sephadex G-100. The active fractions are further purified by absorption to carboxymethyl Sephadex. Purified f_1 is eluted by a suitable concentration of KCl (about 0.3 M) in buffer.

DTT is a stronger reducing agent than 2-mercaptoethanol, and less subject to oxidation by air. It is probably the reducing agent of choice in all buffers (it also smells better). However, it is more expensive so 2-mercaptoethanol is still used, especially where large volumes are involved.

B. Purified f_2

The DEAE-cellulose column (see Section III,A), through which the f_1 passed unbound, is further eluted with a linear gradient of NH_4Cl (0.10-0.35 M) in buffer C. f_3 (see Section III,C) is eluted at about 0.1 M; f_2 elutes at about 0.2 M NH_4Cl (see Figure 1 of Ref. 7). The active fractions containing f_2 are pooled, concentrated by precipitation with $(NH_4)_2SO_4$ (to 90% saturation), and dialyzed overnight against buffer C. This procedure gives an f_2 preparation that still contains contaminating f_1 and f_3, as well as appreciable nuclease and G-factor activity. Several procedures have been described for the further purification of f_2.

Crude f_2 can be freed of contaminating nucleases by passage through a column of hydroxylapatite (Hypatite C,

Clarkson Chemical Company). About 10 mg of crude f_2 in 1 ml of buffer C are applied to a 0.6 X 9-cm column, equilibrated with this buffer, at 0.1 ml/min (a pump is useful). The column is washed successively with 10-ml portions of 0.05 M, 0.075 M, and 0.125 M KH_2PO_4-K_2HPO_4 (pH 7.5) in 0.01 M 2-mercaptoethanol. f_2 is eluted with the 0.125 M buffer, and is recognized as a peak of absorbance at 280 nm. Note that in this procedure, batchwise (as opposed to gradient) elution, a single peak is usually obtained at each salt concentration; all materials adsorbed to the column have apparent R_f values of 0 or 1 in a given eluting buffer. Because of this, the peak that contains f_2 is relatively concentrated. The f_2 activity is further concentrated by precipitation with $(NH_4)_2SO_4$ (90% of saturation), dissolved in buffer [0.01 M tris-HCl (pH 7.8), 0.01 M 2-mercaptoethanol], and dialyzed overnight against this buffer. This preparation is relatively unstable; it loses activity on freezing. It can be kept at 0-4°C for about 1-2 weeks before the activity has decayed to amounts too low to be useful. This preparation of f_2 still contains some T- and G-factor contamination, as well as several of the aminoacyl-tRNA synthetases; it is very low in nuclease activity. In the buffer described f_2 loses over 90% of its activity in 5 min at 50°C.

More recently, several improved procedures for the purification of f_2 have appeared. Herzberg et al. [11] used

6. Chain Initiation Factors from Escherichia coli

essentially the above procedure but introduced a liquid-liquid partition of the crude extract (see Section III,A). They then further purified the f_2 from hydroxylapatite on a DEAE-Sephadex column to give an increase of 2.5-fold in the specific activity for the catalysis of fMet-tRNA binding to poly AUG-ribosome complexes. This preparation of f_2 could be stored for several months in liquid nitrogen in 0.01 M tris-HCl (pH 7.5)-0.07 M 2-mercaptoethanol, or at 0°C in the same buffer in 2 mM $MgCl_2$-50 mM NH_4Cl-50% glycerol.

Kolakofsky et al. described an ingenious technique for the purification of f_2 that takes advantage of its tendency to adsorb to glass surfaces [22]. After initial $(NH_4)_2SO_4$ and DEAE-cellulose steps (substitution of DEAE-Sephadex for DEAE-cellulose has been described [14]), concentrated f_2 (165 mg/14 ml) that had been dialyzed against 20 mM tris-HCl (pH 7.4)-1 mM EDTA-1 mM DTT-20% glycerol was put on a column (15 X 35 cm) of pulverized test tubes (Pyrex; 0.1- to 0.3-mm mesh size) equilibrated with this buffer. After the column was thoroughly washed with the same buffer, f_2 was eluted with a linear gradient of 0-1.5 M NH_4Cl in the buffer. About 75% of the f_2 activity was recovered, in less than 10% of the added protein, at a concentration of about 0.8 M NH_4Cl. The f_2 was concentrated by dialysis against Aquacide II, a high-molecular-weight material that does not pass into a dialysis sac but gives a high enough osmotic pressure to draw water from inside the sac. This purified f_2 preparation was very unstable.

Thach recently described a new method for preparing stable f_2 in high yield and good purity [23]. After an initial $(NH_4)_2SO_4$ fractionation, the crude factors were chromatographed on cellulose phosphate. The initial ribosomal wash was with 0.5 M NH_4Cl; this gave the same yield of f_2 and a higher specific activity (i.e., less removal of other proteins from the ribosomes). The $(NH_4)_2SO_4$ cut was fairly narrow, from 35 to 50% saturation; it gave a slight purification in terms of protein and removed most of the material absorbing at 260 nm (nucleic acid, nucleotides, and so on). 108 mg of protein from the 35-50% fraction was dialyzed against 50 mM tris-HCl (pH 7.4)-20 mM NH_4Cl-1 mM DTT and applied to a 1.2 X 13-cm cellulose phosphate column. A linear gradient of NH_4Cl (20-500 mM) was used to elute f_2; purification was about 36-fold, with recovery of most of the activity applied. The f_2 was concentrated by ultrafiltration (suspension of the material in dialysis tubing in an evacuated flask to remove water by evaporation) and passed through a column of Sephadex G-200 in the above buffer (except that the NH_4Cl concentration was 300 mM) to give a quantitative yield of f_2 and 3-fold further purification. This fraction gives a single band on polyacrylamide gel alectrophoresis (i.e., it is >90% pure).

    This procedure is said to give a 25% recovery of f_2 and a much higher specific activity (S.A. = 47) when the AUG-dependent binding of fMet-tRNA to ribosomes is assayed than the procedure

6. Chain Initiation Factors from Escherichia coli

of Chae et al. (1% recovery, S.A. = 16) outlined below and that of Herzberg et al. (2.2% recovery, S.A. = 3.3) described above.

Chae et al. [12] purified f_2, to a preparation that gave a single band on polyacrylamide gel electrophoresis, by the following sequence of steps: (1) NH_4Cl wash of ribosomes; (2) $(NH_4)_2SO_4$ fractionation; (3) adsorption and elution from $Ca_3(PO_4)_2$ gel; (4) Sephadex G-200 chromatography; (5) DEAE-cellulose chromatography; and (6) hydroxylapatite chromatography. Since the techniques are adequately described in the reference given, no further details are given here. We have not tried this specific procedure; however, in our own experience, adsorption with $Ca_3(PO_4)_2$ gel tends to be tricky and we recommend a pilot-scale run beforehand to ascertain that the specific batch to be used gives the desired results.

Dubnoff and Maitra [19] partially purified f_2 after the DEAE-cellulose step by sequential passage through columns of hydroxylapatite and phosphocellulose.

C. Purified f_3

In the earlier work with initiation factors, only two were known, namely, f_1 and f_2. f_3 was separated only at a later time [7]. As described earlier (Section III), advantage can be taken of the differential solubility of f_2 and f_1 plus f_3 to separate f_3 before the DEAE-cellulose step. Alternatively, the eluate from DEAE-cellulose chromatography of all three factors can be used as the starting material. In either case, f_3 is eluted from DEAE-cellulose at about 0.2 M NH_4Cl (see Section III,A). The active fractions containing f_3 are pooled, dialyzed against 0.1 M phosphate buffer (pH 7.5), and either used as "f_3" or further purified by passage through a phosphocellulose column. After the column is washed with this buffer, active f_3 is eluted with a linear gradient from 0.1 to 0.5 M phosphate buffer (pH 7.5). The peak of activity is pooled and can be further purified by passage through Sephadex G-100; it elutes at a position corresponding to a molecular weight of about 30,000. The f_3 preparation after G-100 chromatography shows four bands on acrylamide gel electrophoresis (one major, three minor) and is several-hundredfold purified over the NH_4Cl wash of the ribosomes; the recovery of the total f_3 activity (MS2 RNA-directed lysine incorporation assay) in the crude extract is about 12% [18]. This f_3 preparation can be further purified by passing it through a column (1.5 X 88 cm) of Sephadex G-75 [24], in a phosphate-tris buffer that contains 6 M urea and DTT, to give homogeneous material (25% yield) that is 1.3-fold purer than the f_3 after the Sephadex G-100 step and is stable for several months at 4°C in 0.5 M phosphate-tris buffer (pH 7.5) containing 10% sucrose. A somewhat simpler purification scheme has been described for the earlier steps [24]. Phosphate is included in all buffers because it stabilizes f_3. Brown and Doty described an uncharacterized

6. Chain Initiation Factors from Escherichia coli

initiation factor that probably corresponds to f_3 [9]. Their purification was as follows. The 1 M NH_4Cl wash from ribosomes was precipitated between 40 and 50% saturation of $(NH_4)_2SO_4$, then chromatographed on a TEAE-cellulose (a variation of DEAE-cellulose) column (1 X 50 cm) with an NH_4Cl gradient. The active material eluted at 0.22 M salt. It was further purified by selective precipitation with $ZnCl_2$. It is still not absolutely certain that this factor is indeed f_3.

The Revel group [8] purified f_3 (factor B) by chromatographing the appropriate fractions from DEAE-cellulose on hydroxylapatite; f_3 was eluted with 0.08 M potassium buffer (pH 7.2). Dubnoff and Maitra [19] used phosphocellulose and carboxymethyl Sephadex.

It seems certain at this time that f_3 corresponds to the ribosomal dissociation factor studied by Davis [25] and Algranati [26]. The identity of f_3 and dissociation factor has been confirmed by direct testing of highly purified f_3 [24,25]. Dissociation factor itself is the crude $(NH_4)_2SO_4$ precipitate of the ribosomal wash from 70 or 30 S ribosomes; it catalyzes the dissociation of 70 S ribosomes into 50 and 30 S subunits.

IV. ASSAYS

The initiation factors may be assayed by virtue of their indispensability for the translation of natural mRNA molecules, or for the binding of certain oligonucleotides to ribosomes, or by measurement of their apparent direct effects on ribosomes.

A. Assays for f_1

(1) Amino acid incorporation into acid-insoluble peptides directed by phage RNA or synthetic oligonucleotides that contain the AUG codon at or near their 5'-end [20]:

This assay is less convenient to perform than the binding assay, and currently is seldom used on a routine basis. It appeared frequently in the early literature.

(2) Binding of fMet-tRNA to ribosomes with the trinucleotide ApUpG as messenger:

This is a rapid, fairly simple assay to perform. The only difficult requirements are a source of f_2 free of f_1, added in limiting quantities, and a source of ribosomes free of initiation factors.

Samples contain, in a final volume of 0.05 ml: 50 mM tris-HCl buffer (pH 7.2); 150 mM NH_4Cl; 3.5 mM magnesium acetate; 0.2 mM GTP; ApUpG, 0.04 A_{260} units; fMet-tRNA labeled with [^{14}C]methionine, 20 pmoles (4800 cpm); purified ribosomes, 2.5 A_{260} units; f_2, about 0.003 units (1 unit causes the binding of 1 nmole of fMet-tRNA in this assay); and glycerol, 2% (v/v). Tubes are incubated for 15 min at 25°C. Incubations are stopped by the addition of 3 ml of ice-cold buffer (0.1 M tris-HCl, 50 mM NH_4Cl, 5 mM magnesium acetate). The sample is filtered as soon as possible through a Millipore filter (HAWP, 0.45-μm pore size) with suction [21]. The filter is washed three times with cold buffer (about 3-5 ml at a time), and sucked as dry as possible. The damp filter is placed in a glass vial (Packard Instruments) or on a planchet and dried in an oven (60-80°C) or under an infrared lamp. The adsorbed radioactivity is determined. Blanks obtained in the absence of f_1 are subtracted from assay values. For liquid scintillation counting a suitable scintillation fluid is Liquifluor (New England Nuclear Corporation, Boston, Massachusetts), a PPO-POPOP mixture in toluene [28]. POPOP is 1,4-bis[2-(4-methyl-5-phenyloxazolyl)]-benzene, a secondary fluor that converts the emitted light from excited PPO to a wavelength suitable for exciting a photomultiplier tube. PPO is 2,5-diphenyloxazole, a primary fluor which emits light of a certain wavelength when excited by an ionizing particle (e.g., the β particle emitted by ^{14}C).

B. Assays for f_2

(1) Binding of fMet-tRNA to ribosomes with the trinucleotide ApUpG as messenger:

This is the same assay as described for f_1, except that the f_1 (about 1 μg; see Ref. 23) is now added in limiting quantities and f_2 is assayed. This is usually the assay of choice as it is simple, quick, and convenient to perform.

(2) Bacteriophage RNA-directed incorporation of amino acids:

This is an assay more commonly used for f_3 and is discussed below (Section IV,C). The procedure can obviously be adapted to assay for f_1 or f_2 by adding the other appropriate factors to the reaction mixture, as all three factors are required for this activity.

(3) f_2-Dependent binding of ribosomes to bacteriophage DNA:

For the investigator with a taste for the elegant and a great deal of leisure time, an assay for f_2-directed binding of ribosomes to mRNA-DNA complexes has been described [8]; an electron microscope is required and the results are not readily quantitated.

(4) Ribosome-coupled GTPase activity:

Purified f_2 has a negligible tendency to hydrolyze the terminal P_i group from GTP. In the presence of purified ribosomes (which have only slight activity alone), however, f_2 has a strong "coupled" GTPase activity, which could serve as the basis for an assay [17]. This procedure is more complicated than the determination of fMet-tRNA binding and is not used on a routine basis.

C. Assays for f_3

(1) Bacteriophage (e.g., MS2) RNA-directed incorporation of amino acids:

This is the assay of choice for f_3 as a suitable routine

6. Chain Initiation Factors from Escherichia coli

binding assay has not been described (however, see Refs. 9 and 24). The assay is convenient for processing large numbers of samples and is fairly simple; again, the difficult requirements are fairly pure f_1 and f_2 free of f_3, ribosomes free of all three factors (washed with 1 M salt solution), and purified phage RNA (see Chapter 11).

Samples contain, in a final volume of 0.125 ml: 60 mM tris-HCl (pH 7.8); 70 mM NH_4Cl; 14 mM magnesium acetate; 16 mM 2-mercaptoethanol; 1.3 mM ATP; 0.3 mM GTP; 17.3 mM phosphocreatine; creatine kinase, 3.1 µg; tRNA (from E. coli), 55 µg; MS2 RNA, 40 µg; purified ribosomes (from an RNase⁻ strain of E. coli, such as Q 13 or MRE 600), 5 A_{260} units; E. coli supernatant (S-150; see Chapter 1) fraction, 0.3 mg of protein; [^{14}C]lysine (10 Ci/mole); 0.1 mM (each) mixture of 19 unlabeled amino acids; f_1 (50% pure), about 2 µg; f_2 (10% pure), about 40 µg. Tubes are incubated for 20 min at 37°C. Incubations are stopped by the addition of 5% trichloroacetic acid (TCA) (to precipitate proteins and RNA), then incubated for 15 min at 90°C (to hydrolyze and dissolve RNA). The remaining precipitate after cooling (protein) is collected on Millipore filters and washed with cold 5% TCA on the filter (see Chapter 1). The filters are placed in vials, dried, and counted in a liquid scintillation counter after the addition of a suitable solvent (for example, Bray's solution [27]). Blank values (in the absence of f_3) are subtracted from all determinations.

(2) Other assays:

Other assays described for f_3 include an assay based on its effect on lysozyme synthesis directed by bacteriophage T_4 in vitro [8], its requirement for the binding of natural mRNAs to Millipore filters [9,18], its requirement for fMet-tRNA binding to Millipore filters (with highly purified factors [24]), and its requirement for the dissociation of 70 S ribosomes to 30 and 50 S ribosomal subunits [24,25].

REFERENCES

[1]   W. M. Stanley, Jr., M. Salas, A. J. Wahba, and S. Ochoa, Proc. Natl. Acad. Sci. U.S., 56, 290 (1966).

[2]   G. Brawerman and J. M. Eisenstadt, Biochemistry, 5, 2777 and 2784 (1966).

[3]   M. Revel and F. Gros, Biochem. Biophys. Res. Commun., 25, 124 (1966).

[4]   H. F. Lodish, Nature, 226, 705 (1970).

[5]   D. A. Shafritz and W. F. Anderson, J. Biol. Chem., 245, 5553 (1970).

[6]   S. M. Heywood, Proc. Natl. Acad. Sci. U.S., 67, 1782 (1970).

[7]   K. Iwasaki, S. Sabol, A. J. Wahba, and S. Ochoa, Arch. Biochem. Biophys., 125, 542 (1968).

[8]   M. Revel, M. Herzberg, and H. Greenshpan, Cold Spring Harbor Symp. Quant. Biol., 34, 261 (1969).

6. Chain Initiation Factors from Escherichia coli

[9] J. C. Brown and P. Doty, Biochem. Biophys. Res. Commun., 30, 284 (1968).

[10] J. M. Eisenstadt and G. Brawerman, Proc. Natl. Acad. Sci. U.S., 58, 1560 (1967).

[11] M. Herzberg, J. C. Lelong, and M. Revel, J. Mol. Biol., 44, 297 (1969).

[12] Y. B. Chae, R. Mazumder, and S. Ochoa, Proc. Natl. Acad. Sci. U.S., 62, 1181 (1969).

[13] M. Salas, A. M. Smith, W. M. Stanley, Jr., A. J. Wahba, and S. Ochoa, J. Biol. Chem., 240, 3988 (1965).

[14] T. Ohta and R. E. Thach, Nature, 219, 238 (1968).

[15] M. Revel, G. Brawerman, J. C. Lelong, and F. Gros, Nature, 219, 1016 (1968).

[16] J. W. B. Hershey, K. F. Dewey, and R. E. Thach, Nature, 222, 944 (1969).

[17] R. E. Thach, J. W. B. Hershey, D. Kolakofsky, K. F. Dewey, and E. Remold-O'Donnell, Cold Spring Harbor Symp. Quant. Biol., 34, 277 (1969).

[18] A. J. Wahba, Y. B. Chae, K. Iwasaki, R. Mazumder, M. J. Miller, S. Sabol, and M. G. Sillero, Cold Spring Harbor Symp. Quant. Biol., 34, 285 (1969).

[19] J. S. Dubnoff and U. Maitra, Cold Spring Harbor Symp. Quant. Biol., 34, 301 (1969).

[20] M. Salas, M. B. Hille, J. A. Last, A. J. Wahba, and S. Ochoa, Proc. Natl. Acad. Sci. U.S., 57, 387 (1967).

[21] M. Nirenberg and P. Leder, Science, 145, 1399 (1964).

[22] D. Kolakofsky, K. F. Dewey, and R. E. Thach, Nature, 223, 694 (1969).

[23] E. Remold-O'Donnell and R. E. Thach, J. Biol. Chem., 245, 5737 (1970).

[24] S. Sabol, M. G. Sillero, K. Iwasaki, and S. Ochoa, Nature, 228, 1269 (1970).

[25] A. R. Subramanian and B. D. Davis, Nature, 228 1273 (1970).

[26] I. D. Algranati, N. S. Gonzalez, and E. G. Bade, Proc. Natl. Acad. Sci. U.S., 62, 574 (1969).

[27] G. A. Bray, Anal. Biochem., 1, 279 (1960).

[28] J. A. Last, W. M. Stanley, Jr., M. Salas, M. B. Hille, A. J. Wahba, and S. Ochoa, Proc. Natl. Acad. Sci. U.S., 57, 1062 (1967).

Chapter 7

CHAIN ELONGATION FACTORS

Julian Gordon

The Rockefeller University
New York, New York

I. INTRODUCTION 177
II. GENERAL FEATURES OF THE PREPARATION. 179
III. GROWTH CONDITIONS AND PREPARATION OF EXTRACTS. . . . 181
IV. SEPARATION OF T AND G FACTORS. 188
V. SEPARATION OF T_U AND T_S. 192
VI. RECOVERY OF THE RIBOSOMES. 192
VII. ASSAYS 194
 REFERENCES 197

I. INTRODUCTION

The polypeptide chain elongation factors were first isolated from mammalian sources, required, in addition to ribosomes, to carry out the continued synthesis of already initiated nascent peptide chains primed by endogenous mRNA [1, 2]. These factors were later found to be required for polyphenylalanine synthesis primed by exogenous poly U [3]. A comparable pair of factors was then isolated from bacterial systems [4]. It was not realized at the time that unphysio-

Copyright © 1971 by Marcel Dekker, Inc. No part of this
work may be reproduced or utilized in any form or by any means,
electronic or mechanical, including xerography, photocopying,
microfilm, and recording, or by any information storage and
retrieval system, without the written permission of the publisher.

logically high concentrations of magnesium circumvented the requirement for specific initiator codons, tRNA, and initiation factors (see Chapter 3). Subsequently, one of the bacterial factors, T, was found to be resolvable into two subfactors, T_U and T_S [5]. However, these two tend to remain in association with each other and for many purposes can be regarded as a single factor.

Several groups have been working on the elongation factors, and a variety of terminology has been used. Table 1 summarizes the terminology and the species from which these factors were

Table 1

Sources and Terminology of Polypeptide Chain Elongation Factors

	Type of factor		
Source	Aminoacyl-tRNA binding	Translocase	Reference
Escherichia coli	T	G	[6]
Escherichia coli	T	G	[7]
Escherichia coli	T	G	[8]
Escherichia coli	--	G	[9]
Escherichia coli	Fl_U, Fl_S	F2	[10]
Escherichia coli	T_U, T_S	--	[11]
Escherichia coli	--	G	[12]
Pseudomonas fluorescens	T_U, T_S	G	[5]
Bacillus stearothermophilus	S3, S1	S2	[13]
Yeast cytoplasm	T_{cyt}	T_{cyt}	[14]
Yeast mitochondria	T_{mit}	G_{mit}	--
Rabbit reticulocytes	TF-1	TF-2	[15]
Rat liver	Aminoacyl transferase I	Aminoacyl transferase II	[16]

7. Chain Elongation Factors

isolated. The references are to the key paper, where available, for the method of preparation used by a group.

In spite of the variety in terminology, there is a uniformity in functional characteristics that justifies classifying them as "aminoacyl-tRNA binding" or "translocase" as shown in the table. The binding factors stimulate the binding of aminoacyl-tRNA to ribosomes, take part in the formation of an intermediary complex with aminoacyl-tRNA and GTP, and have GTPase activity that is dependent on both aminoacyl-tRNA and ribosomes. The translocases share the following properties: they stimulate the translocation of peptidyl-tRNA from one site to another on the ribosome and are characterized by the ability to hydrolyze GTP only in the presence of ribosomes. References to these aspects can be found throughout the literature cited in Table 1, as well as reviews in Refs. 17 and 18. The similarity of the factors from various species is also emphasized by the interchangeability between one species and another [19]. All this suggests a homologous series, derived from a common ancestor at some time in the remote past. It is hoped that a common terminology will be adopted in the not-too-distant future. In this chapter the T (or T_U, T_S) and G terminology is retained.

II. GENERAL FEATURES OF THE PREPARATION

The methodology is based mainly on that described in

Ref. 8, as well as unpublished alternatives which lend themselves better to larger- or smaller-scale preparations.

The secret of a good preparation is careful attention to the following three points: (1) attention to growth conditions; (2) attention to the conditions of storage of the cells; (3) the use of dithiothreitol (DTT) as a sulfhydryl (SH) group protector. It is very easy to ignore (1) and (2) and to regard a frozen package of cells as the starting point of the preparation. In the earlier preparations, not much attention was paid to these points, and DTT was not available. The factors therefore acquired the reputation for being impossible to work with. For example, when 2-mercaptoethanol was used as an SH protector, the half-life of T was approximately 1 day. The replacement of 2-mercaptoethanol by DTT extended the half-life to 6 weeks at $0°C$. The use of liquid nitrogen storage prolongs the life indefinitely. (The use of open tubes is to be recommended for the liquid nitrogen storage; this both eliminates the danger of explosion and provides an oxygen-free atmosphere.)

The point concerning oxygen should be emphasized. Both 2-mercaptoethanol and glutathione are oxidizable by molecular oxygen. It is not advisable to use them in combination with DTT as they are able to act as substrate for the oxidation of the DTT. All the oxidation products act as irreversible inactivating agents especially for T factor. However, DTT should not be regarded as absolutely stable. It seems to be

7. Chain Elongation Factors

least stable in dilute solutions, and therefore buffers and columns should not be allowed to stand around unnecessarily when DTT is present. Stock solutions at 1 M concentration seem to be quite stable in a freezer. As for most enzyme preparations, all operations are carried out in an ice bucket at 0°C wherever possible.

III. GROWTH CONDITIONS AND PREPARATION OF EXTRACTS

The reason for attention to growth rate is that as the cells grow faster a higher concentration of elongation factors (as well as ribosomes) is needed to maintain their rate of growth [20]. Almost any rich medium can be used, and the one used here reliably and reproducibly results in a generation time of 24 min up to high cell densities (3 X 10^9 cells per milliliter). The medium consists of the following basic salts mixture: $MgSO_4 \cdot 7H_2O$, 0.1 g/liter; citric acid, 1.0 g/liter; $Na_2HPO_4 \cdot 2H_2O$, 5.0 g/liter; $Na(NH_4)HPO_4$, 1.74 g/liter; KCl, 0.74 g/liter; the pH is 7.0. This mixture can be stored as a 50-fold concentrate. The medium is supplemented with 0.5% Difco nutrient broth, 0.8% Difco yeast extract, and 1.0% glucose. Glucose is best sterilized separately, and we routinely guard against contamination with spores by autoclaving the medium twice, 15 min each time.

Shaking or bubbling is inadequate to obtain good growth up to the desired cell densities. It is best to use stirring, with as large a surface area as possible for the stirrer blades, in a 15-liter carboy. This provides a vortexlike motion, and the aeration is adjusted so as to just prevent excessive foaming. Antifoam should be avoided. The result is a fine dispersion of bubbles throughout the whole medium.

The inoculum is a 1/100 dilution of an overnight shake-flask culture in the same medium, and the growth is followed by the turbidity of the culture. [It should be pointed out that the strain does not make any difference, provided it grows in this medium reasonably rapidly. We have usually used E. coli strain S/6 (a derivative of B), but K12 and C have also been used. In fact, the T and G factors of a number of closely related species (Salmonella typhimurium, Shigella dysenteriae, Enterobacter aerogenes) behave indistinguishably from those of E. coli, at least in polyacrylamide electrophoresis.] Any spectrophotometer or colorimeter can be used; we usually use a Zeiss instrument at a wavelength of 660 nm. The cells are harvested when the absorbance reaches 3.0. Growth is stopped by pouring the culture onto ice. A 15-liter culture provides approximately 25 g of wet cell paste after harvest.

For this scale it is most convenient to harvest the cells in a Szent-Györgi-Blum continuous-flow attachment for a Servall refrigerated centrifuge. The relatively small amount of cells is conveniently collected in stainless steel centrifuge tubes. The Sharples model is better suited for collecting larger

7. Chain Elongation Factors 183

amounts of cells, and it is too tedious to use a large-capacity angle rotor. If such a continuous-flow system is not available, it is possible to separate the cells by filtration with the aid of celite (Hyflo-Super-Cel, Amend Corporation; also known as kieselguhr or diatomaceous earth). First, a bed is laid down by suspending 25 g of celite in 0.1 M NH_4Cl-10 mM tris-HCl (pH 7.8)-10 mM $MgCl_2$ and pouring the mixture over filter paper in a large Büchner funnel (18-in. diameter or more). This is partially filtered but not allowed to run dry. The chilled culture is then mixed with 50 g of celite and rapidly filtered on the funnel. The timing must be such that the celite does not settle down in the middle of the filtration, otherwise the cells clog the filter. If there is a tendency for this to happen, it can be prevented by stirring the suspension during the filtration. The cells and celite together form a cake which can be peeled off the filter paper and ground in a mortar and pestle in the usual way. The celite can substitute for alumina in the grinding step.

The harvested cells can be fast-frozen in liquid nitrogen and stored. However, the factors are quite stable at 0°C and can be kept overnight on ice with no loss in activity. Storage at -20°C in a conventional freezer should be avoided.

Any one of several methods can be used for homogenizing the cells. As mentioned above, grinding with a mortar and pestle may be used, and this is described in detail in Chapter

3. Blending with glass beads or passing through a small orifice at high pressure are also effective methods. Ultrasonic treatment seems to result in a considerable loss of activity. The most desirable method is determined by the scale of the operation. For a very small scale, the method of Godson and Sinsheimer is most suitable [21]. This lysozyme-nonionic detergent procedure enables one to control the lytic process in such a way as to leach out the cytoplasm selectively, leaving the polysomes and DNA inside the cell. The T and G factors are released quantitatively, and there is no need for high-speed centrifugation. This method is very easily reproduced from the directions in the original publication, so no details are given here. For a 10- to 50-g scale, a French press is most suitable (as manufactured by Aminco and described in Ref. 22). For 500 g of cells or more, a Manton-Gaulin mill is most suitable (Manton-Gaulin Manufacturing Company). This mill works in principle similarly to a French press, except that the suspension is pumped continuously through an orifice under high pressure.

For breakage of cells in a French press, the cell paste (25 g) is mixed with a minimal volume of buffer A (10 mM tris-HCl (pH 7.8), 10 mM $MgCl_2$, 1 mM DTT) just sufficient to render the suspension fluid. The buffer is added a little at a time to prevent lumps from forming and to maintain a uniform consistency. DNase (50 μg, Worthington, electrophoretically

7. Chain Elongation Factors 185

purified) is mixed in, and the suspension is passed through the precooled press at approximately 10,000 psi. The DNase works instantaneously, and there is no need for any incubation. If for any reason the DNase fails to work or is omitted, this concentration of cells results in an intractable glop. Otherwise, the homogenate is just as fluid as the original cell suspension. The homogenate is then taken up in 50 ml of buffer and the cell debris removed as follows. The homogenate is centrifuged at 30,000 X g for 30 min and the supernatant carefully decanted and put aside. The pellet is extracted with an additional 25 ml of buffer A and centrifuged a second time. The supernatant is again decanted and combined with the previous one. The pooled supernatants are then completely clarified by another 30 min centrifugation at 30,000 X g. This fraction is referred to as the S-30.

For $(NH_4)_2SO_4$ fractionation to work well, it is vital that all nucleic acids and ribosomes be quantitatively removed from the preparation. If this is done, then the T and G factors will be separated in a clean, simple batchwise procedure (see Section IV). There are two approaches to achieving this end. The more conventional one (1) is to pellet the ribosomes by high-speed centrifugation and then to precipitate out the remaining nucleic acid with protamine sulfate. The other (2) has been developed because of its speed, simplicity, and ease of scaling up.

Alternative (1): A conventional S-100 is made by centrifuging S-30 at 105,000 X g for 90 min, and the top 10 ml of each tube is aspirated (13-ml tubes of a Spinco 40 rotor). The tRNA is then precipitated by the addition of an amount of protamine sulfate just sufficient to titrate the phosphate groups of the tRNA. Protamine sulfate is extremely variable, both from supplier to supplier and from batch to batch, so each lot must be individually calibrated. This is easily done by following the disappearance of the 260-nm absorbance of tRNA in S-100. Protamine sulfate is most conveniently made up as a 1% stock solution, neutralized with KOH, and the insoluble material centrifuged out. Typically, 1/4 vol suffices to precipitate out the tRNA quantitatively. The mixture is stirred for 10 min, and the precipitate is removed by centrifugation. The T and G remain in the supernatant. These are then precipitated by the addition of 0.455 g of $(NH_4)_2SO_4$ for each milliliter of S-100 with stirring; the stirring is continued for 30 min. The precipitate is removed by centrifugation and the supernatant is processed as described in Section IV.

Alternative (2): The system used here is the polyethylene glycol-dextran aqueous two-phase partition system [23]. The ribosomes and tRNA partition into the bottom (dextran) phase, and the factor proteins are mostly partitioned into the top (polyethylene glycol) phase. The salt concentration determines

7. Chain Elongation Factors

the partition coefficient of any given protein, and Na^+ is the best ion to use for this purpose. However, NH_4^+ has been selected for our work as it permits the recovery of active ribosomes from the bottom phase. The following ingredients are mixed, with stirring, in the order given: S-30 fraction, 80 ml; polyethylene glycol 6000 (30% w/w), 22.5 ml; dextran 500 (20% w/w), 7.5 ml; solid NH_4Cl, 11.3 g. The solutions are expressed as w/w because the high viscosity and the tendency to foam make it difficult to bring them up to a constant volume, and for the same reason the volumes are measured and dispensed with syringes (not needles). We have found one batch of polyethylene glycol (Ruger Chemical Corporation) that failed to form the two-phase system, and this has also been reported [23] for one batch of Dextran (Pharmacia). The phase system should therefore be tested with water beforehand in order to avoid sacrificing the material.

The mixture is stirred for 10 min, and the phases separated by centrifugation for 10 min at 2,000 X g. The viscosity of the bottom phase is so high that the top phase can be poured off without any trouble. The pH of the top phase is then brought to 7.8 by the addition of NH_4OH. After this has been done, the pH remains stable during the subsequent additions of $(NH_4)_2SO_4$. The next step is to remove the polyethylene glycol. This is achieved in a second partition system. Polyethylene glycol and $(NH_4)_2SO_4$ form such a system [24].

Solid $(NH_4)_2SO_4$ (16.4 g) is added with stirring, and the phases are again separated by 10-min centrifugation at 2000 g. There are two phases; the small top phase is separated from the bottom phase by an interface of precipitated protein. The bottom [$(NH_4)_2SO_4$] phase contains the factors. This is the trickiest step: the bottom phase is removed by aspiration by inserting a thin plastic tube through the interface. The T and G factors are then precipitated by the addition of $(NH_4)_2SO_4$ to saturation (approximately 20 g); the solution is stirred for 30 min. The precipitate is removed by centrifugation and the fractionation is continued as follows.

IV. SEPARATION OF T AND G FACTORS

Separation of T and G factors is accomplished by successive extraction of the $(NH_4)_2SO_4$ precipitates with decreasing concentrations of $(NH_4)_2SO_4$. We define the ammonium concentrations as weight percent; this is unconventional but operationally unambiguous (i.e., 27.8% is 27.8 g + 72.2 g of water). Solutions of 27.8, 24.0, and 19.6% are made up, adjusted to pH 7.8 with concentrated NH_4OH, and then brought to 1 mM DTT. These solutions correspond approximately to 55, 45, and 35% saturation as conventionally expressed. The precipitated factors are extracted successively with 7.5 ml of each of these solutions, each extraction being repeated three times. At 24.0% G factor is completely insoluble, and T factor

7. Chain Elongation Factors

is partially soluble. About 70% of the T factor is obtained free of G in these fractions. Beyond this point the T and G are processed separately but in similar steps. The 24.0% eluates are pooled, as are the 19.6% eluates. The T and G factors are concentrated from these by reprecipitation with 4.5 g of $(NH_4)_2SO_4$. The precipitates are collected by centrifugation and are resuspended in about 1 ml of buffer B [10 mM tris-HCl (pH 7.8), 1 mM DTT]. The material is then transferred to a dialysis sac and dialyzed against 500 ml of buffer B. The buffer is changed twice over a period of about 12 hr. Thereafter, the processing of the T and G factors may be either in parallel, or either one may be put aside in liquid nitrogen for later use. However, either or both is further purified on DEAE-cellulose and then on hydroxylapatite columns.

DEAE-cellulose column chromatography is performed as follows. T factor is purified on a column 15 X 0.6 cm and G factor is purified on a column that is 20 X 0.8 cm. These are designed to be approximately 30-50% of maximal loading of the column. While low loadings theoretically give higher resolution, they lead to low recoveries of activity. The columns are packed with DEAE-cellulose (Biorad, exhaustively washed with acid and alkali) equilibrated with 0.1 M KCl-10 mM tris-HCl (pH 7.8). DTT is run into the column only a few minutes before applying the sample. It is a good idea to check routinely the conductivity of the sample before applying it to the column, in case the dialysis has been insufficient. Ensuring that the conductivity of the sample is at least an order of magnitude below that of the starting buffer will avoid a good deal of disappointment.

The sample is applied to the column by layering it beneath the starting buffer (the very concentrated protein solution should be denser than the starting buffer). This eliminates the need for repetitive washing of the top of the column, and the dense band enters the column cleanly. The column is washed through with the starting buffer (buffer B and 0.1 M KCl) until the breakthrough protein peak is completely eluted. The protein is measured at 235 nm, and an absorbance of 3 corresponds very closely to 1 mg/ml. The factor is then eluted from the column with a linear salt gradient from 0.1 M KCl to 0.3 M KCl in buffer B, with 100 ml in each reservoir. The peak of the factor activity is then located by assay. This is actually the only point in the preparation where it is necessary to interrupt the procedure to locate the active fractions.

The tubes containing the activity are pooled and the protein is concentrated by $(NH_4)_2SO_4$ precipitation. The precipitation is carried out by the addition of 0.455 g of $(NH_4)_2SO_4$ for each milliliter of pooled material. The solution is stirred for 30 min; the precipitate is centrifuged off, resuspended in a minimal volume of buffer B, and dialyzed against two changes of buffer B.

7. Chain Elongation Factors

The material is then fractionated by stepwise elution from hydroxylapatite. Batches of hydroxylapatite are very variable, both in the flow characteristics of the hydroxylapatite and in the salt concentration at which a given protein is adsorbed. It is therefore a good idea to buy a large quantity and thus avoid the necessity of intermittently checking the characteristics of different batches. The hydroxylapatite (we use Hypatite C from the Clarkson Chemical Company) is packed into a 0.5 X 3.0 cm column and equilibrated with 10 mM KH_2PO_4-K_2HPO_4 buffer (pH 7.2)-1 mM DTT, again adding the DTT only at the last minute. It is not necessary to prewash the hydroxylapatite in any way. The active fraction from DEAE-cellulose (about 15 mg of protein) is applied to the column in the same way as for the DEAE-cellulose step; the column is washed successively with 50 ml of 30 and 60 mM phosphate buffer in the case of T factor, and with 10 and 30 mM buffer in the case of G factor. T-Factor activity elutes coincident with the protein 235-nm peak in the 60 mM eluate, and G-factor activity elutes coincident with the protein in the 30 mM eluate.

The factors are about 70% pure at this stage and do not contain any detectable cross-contamination. They are concentrated by $(NH_4)_2SO_4$ precipitation as described for the DEAE fractions and either stored in liquid nitrogen after dialysis, or the T factor is split into its constituent T_U and T_S components.

V. SEPARATION OF T_U AND T_S

T Factor is separated into its constituent subcomponents by the dissociation of T in the presence of Mg^{2+} and GDP [25] (unpublished procedure). A very small DEAE-cellulose column is made in a Pasteur pipet and equilibrated as before. T Factor (about 10 mg of protein) is applied to the column which is washed with a few milliliters of 0.1 M KCl in buffer B. T_S is then specifically eluted by washing the column with the same buffer, containing 10 mM $MgCl_2$ and 25 µM GDP. At this salt concentration GDP and T_S do not bind to the column, but T_U does. T_U is then eluted by raising the salt concentration to 0.15 M KCl. The activities of the separated T_U and T_S can then be assayed, as they are both needed for the binding of additional radioactive GDP and neither T factor alone binds it. Actually, the GDP bound is exchanged for the endogenously bound GDP when [^{3}H]GDP is added, a reaction catalyzed by T_S with T_U acting as the acceptor [25].

VI. RECOVERY OF THE RIBOSOMES

Ribosomes can be prepared by any one of a variety of methods, the most usual being about five washes in 0.5-1.0 M NH_4Cl. The procedure described in the chapter on initiation can be used, but four or five washes are needed before an absolute requirement for G is obtained. The ribosome pellet from alternative (1) (Section III) is used as the starting

7. Chain Elongation Factors

material. Alternatively, the ribosomes can be recovered from the dextran phase in alternative (2) (Section III). Recovery from the dextran phase is described in detail. There are two main steps; first the dextran is removed by $(NH_4)_2SO_4$ fractionation; then the contaminating factors are separated on a Sephadex G-200 column.

The dextran phase (see Section III) is diluted to a final volume of 200 ml in 10 mM tris-HCl (pH 7.8)-10 mM $MgCl_2$-0.1 mM DTT (buffer C), and 42 g of $(NH_4)_2SO_4$ are added with stirring for 10 min. The precipitate is then removed by centrifugation, and the procedure is repeated. The pellet contains the ribosomes, which are then resuspended in buffer C and dialyzed against 1 liter of the same buffer. The material is next passed through a 2 X 40 cm Sephadex G-200 column equilibrated with buffer C in 1 M NH_4Cl. The ribosomes appear in the void volume and are located by their absorbance at 260 nm. As with all Sephadex G-200 columns of any size, care should be taken to avoid subjecting the column to any hydrostatic pressure while it is flowing. Otherwise, very poor flow rates are obtained. The pressure is therefore controlled with a constant-head device (Mariotte flask), at approximately 20 cm of water. The pooled ribosome peak (including fractions up to the point where the absorbancy has dropped to about 50% of the peak height) is then centrifuged to pellets and concentrates the ribosomes (4.5 hr at 165,000 g). The pellets are resuspended in buffer C at 1440 A_{260} units per milliliter (100 mg/ml) and stored at −16°C. This procedure has the advantage that there is no selective loss of the small subunit such as occurs during repetitive centrifugation, and it also avoids the tedium of more conventional procedures.

VII. ASSAYS

The poly-U system is the most foolproof way of assaying both factors, although the reaction requires more ingredients than the alternatives. The reaction mixtures contain in 0.125 ml: 50 mM tris-HCl (pH 7.4); 160 mM NH_4Cl; 10 mM $MgCl_2$; 12 mM DTT; poly U, 5 µg; Phe-tRNA, 50 µg; (25 pmoles, either ^{14}C or ^{3}H), ribosomes, 50 µg; and 125 nmoles GTP. The T and G factors are added at concentrations that are adjusted to be limiting or in excess, depending on which factor is to be measured. To begin, 5 µl of a 10-fold dilution of the S-100 fraction (i.e., 0.5 µl) saturates this reaction. All the ingredients except for tRNA, ribosomes, GTP, and factors can be stored as a 2.5 X concentrated stock solution which can be kept for several months in a freezer. All the components of the reaction mixture except GTP are mixed together, and the reaction is started by the addition of GTP. Incubation is for 10 min at 30°C, and the reaction is stopped by the addition of about 2 ml of 5% trichloroacetic acid (TCA). The unreacted Phe-tRNA is hydrolyzed by heating for 10 min at 90-95°C. The

7. Chain Elongation Factors

hot TCA-insoluble radioactivity is collected on a Millipore filter, washed four times with 5 ml (each time) of 5% TCA, dried, and counted. Counting conditions are the same as described in Chapter 3. The volumes used in the filtration are not critical, so it is perfectly satisfactory to squirt the TCA with a wash bottle. This saves a great deal of time if there are many assays involved. The Phe-tRNA is prepared by any one of several procedures, and references can be found throughout the literature cited. Poly U can be obtained from Miles.

If it is desired to assay the factors separately, by assay methods that are independent of the availability of both T and G, then GDP binding is the most suitable assay for T factor, while the ribosome-dependent GTPase is the best assay for G factor.

The reliability of the GDP binding assay stems from the very high affinity of T factor for GDP [26]. Reaction mixtures contain the same ingredients as the poly-U system above, except that poly U, tRNA, ribosomes, and GTP are not added. Instead, 400 pmoles of [^{3}H]GDP are added. Up to 200 μg of protein can be present in the assay. More than this overloads the Millipore filter. All the reagents except T factor are stored as a 2.5 X concentrated mixture, and the concentrated stock solution is filtered through a Millipore filter beforehand in order to decrease the blanks. Reactions are started by adding the enzyme and are run for 30 sec at 0°C. They are then stopped by the addition of 5 ml of cold 10 mM tris-HCl (pH 7.8)-10 mM $MgCl_2$. The mixture is immediately poured over a Millipore filter and washed three times with 5 ml (each) of the same buffer. The filter which is not allowed to run dry between washings, is dried and counted as described above. The wash buffer should also be passed through a Millipore filter beforehand.

The GTPase assay for G factor is carried out as follows. The reaction mixtures are the same as for the poly-U system, except that poly U, tRNA, and GTP are omitted. Instead, 2.5 nmoles of [γ-^{32}P]GTP (International Chemicals and Nuclear Company) are present. Again, a 2.5 X concentrated mixture of the salts and DTT is used as a stock solution which is stable in a freezer. The ^{32}P is diluted with cold GTP to give about 20,000 cpm per reaction mixture. The $^{32}P_i$ released can be measured either by charcoal adsorption of the unreacted [^{32}P]GTP, or by extraction of the $^{32}P_i$ as a phosphomolybdate complex in an organic solvent. The latter procedure has proved most useful to us as it involves the collection of the top phase from a two-phase system. Collecting and counting filtrates is always more cumbersome. However, it should be pointed out that the phosphomolybdate is impossible to count in a scintillation counter, so the method chosen should ultimately depend on the counting system available. The phosphomolybdate extraction is described in detail.

7. Chain Elongation Factors

The reaction is started by the addition of [^{32}P]GTP, run for 10 min at 30°C, and stopped by the addition of 0.5 ml of cold 0.02 M silicotungstate in 0.02 N H_2SO_4. The entire procedure is run in the cold. Then, 1.25 ml of carrier, P_i and 0.5 ml of 5% ammonium molybdate in 4 N H_2SO_4 are added, followed by 2.5 ml of isobutanol-benzene (1:1, v/v). The mixture is agitated in a vortex mixer for about 30 sec and the phases are separated by 5 min of low-speed centrifugation. The top phase is transferred to a deep planchet and dried under an infrared lamp. If very precise quantitation is required, the bottom phase can be reextracted with 1.5 ml of the same organic solvents. The material is then counted in a gas-flow counter.

ACKNOWLEDGMENT

The unpublished experiments discussed here, as well as the published parts of my work that have been cited, were all performed in the laboratory of Dr. Fritz Lipmann under Grant no. G.M.13972 from the National Institutes of Health. Dr. Lipmann's critical and stimulating discussions throughout are gratefully acknowledged.

REFERENCES

[1] J. O. Bishop and R. S. Schweet, Biochim. Biophys. Acta, 54, 617 (1961).

[2] J. M. Fessenden and K. Moldave, Biochem. Biophys. Res. Commun., 6, 232 (1961).

[3] R. Arlinghaus, G. Favelukes, and R. Schweet, Biochem. Biophys. Res. Commun., 11, 92 (1963).

[4] J. E. Allende, R. Monro, and F. Lipmann, Proc. Natl. Acad. Sci. U.S., 51, 1211 (1964).

[5] J. Lucas-Lenard and F. Lipmann, Proc. Natl. Acad. Sci. U.S., 55, 1562 (1966).

[6] Y. Nishizuka and F. Lipmann, Proc. Natl. Acad. Sci. U.S., 55, 212 (1966).

[7] A. Parmeggiani, Biochem. Biophys. Res. Commun., 30, 613 (1968).

[8] J. Gordon, J. Biol. Chem., 244, 5680 (1968).

[9] P. Leder, L. E. Skogerson, and M. M. Nau, Proc. Natl. Acad. Sci. U.S., 62, 454 (1969).

[10] R. L. Shorey, J. M. Ravel, C. W. Garner, and W. Shive, J. Biol. Chem., 244, 4555 (1969).

[11] J. M. Ravel, Proc. Natl. Acad. Sci. U.S., 57, 1811 (1966).

[12] R. Ertel, N. Brot, B. Redfield, J. E. Allende, and H. Weissbach, Proc. Natl. Acad. Sci. U.S., 59, 861 (1968).

[13] A. Skoultchi, Y. Ono, H. M. Moon, and P. Lengyel, Proc. Natl. Acad. Sci. U.S., 60, 675 (1968).

[14] D. Richter and F. Lipmann, Biochemistry, 9, 5065 (1970).

[15] W. L. McKeehan and B. Hardesty, J. Biol. Chem., 244, 4330 (1969).

[16] W. Galasinski and K. Moldave, J. Biol. Chem., 244, 6527 (1969).

[17] P. Lengyel and D. Söll, Bacteriol. Rev., 33, 264 (1969).

[18] J. Lenard and F. Lipmann, Ann. Rev. Biochem., in press.

[19] I. Krisko, J. Gordon, and F. Lipmann, J. Biol. Chem., 244, 6117 (1969).

[20] J. Gordon, Biochemistry, 9, 912 (1970).

[21] G. N. Godson and R. L. Sinsheimer, Biochim. Biophys. Acta, 149, 476 (1967).

[22] C. S. French and H. W. Milner, in Methods in Enzymology, Vol. 1 (S. P. Colowick and N. O. Kaplan, eds.), Academic Press, New York, 1955, p. 64.

[23] B. M. Alberts, in Methods in Enzymology, Vol. 12A (L. Grossman and K. Moldave, eds.), Academic Press, New York, 1967, p. 566.

[24] P. -A. Albertsson, Partition of Cells and Macromolecules, John Wiley, New York, 1960.

[25] H. Weissbach, D. L. Miller, and J. Hachmann, Arch. Biochem. Biophys., 137, 262 (1970).

[26] D. Cooper and J. Gordon, Biochemistry, 8, 4289 (1969).

Chapter 8

PREPARATION OF POLYPEPTIDE TERMINATION FACTORS

FROM Escherichia coli

Edward M. Scolnick

Viral Lymphoma and Leukemia Branch
National Cancer Institute
Bethesda, Maryland

I. INTRODUCTION 201

II. PREPARATION OF MATERIALS 203

   A. Escherichia coli. 203

   B. Ribosomes 203

   C. Codons. 204

   D. f[^{3}H]Met-tRNA^f. 204

III. f[^{3}H]MET-tRNA-AUG-RIBOSOME COMPLEX 206

IV. RELEASE ASSAY. 207

V. PREPARATION OF R1 AND R2 208

   A. Other Assays. 211

   REFERENCES 211

I. INTRODUCTION

Polypeptide chain termination may be defined as the release of nascent peptide chains from ribosomes upon translation of one of the three terminator codons, UAA, UAG, or UGA. The mechanism of release requires a minimum of two events: (1)

recognition of terminator codons, and (2) hydrolysis of the ester linkage between the growing peptide chain and the tRNA by which it is bound to the ribosome. Capecchi [1] was the first to describe a release factor from fractions from E. coli B, S-100 which was necessary for the release of nascent peptide chains upon translation of the codon UAG. His assay employed an mRNA from a mutant of bacteriophage R17 in which the seventh codon in the viral coat protein cistron had mutated from CAG to UAG. In a cell-free protein-synthesizing system, this mutant mRNA directs the synthesis of the hexapeptide fMet-Ala-Ser-Asn-Phe-Thr, which is released upon translation of the seventh codon, UAG. He showed that the termination factor necessary for the release of this hexapeptide was a protein and that it caused the release event to occur only at the UAG codon position of the message, not at the threonine position (ACC).

A second assay developed to study termination involved the use of trinucleotide codons and bypassed the intermediate elongation steps of protein synthesis. Caskey et al. [2] described an assay in which trinucleotide codons are used in a sequential manner to bind N-formyl[^{3}H]methionyl-tRNA (f[^{3}H]Met-tRNA), with AUG as a messenger to E. coli B ribosomes and subsequently to release f[^{3}H]methionine with terminator codons and release factor:

8. Preparation of Polypeptide Termination Factors

f[^{3}H]Met-tRNA + ribosomes + AUG $\rightleftharpoons$ f[^{3}H]Met-tRNA-AUG-ribosome

f[^{3}H]Met-tRNA-AUG-ribosome + R factor + terminator codon
$\rightarrow$ f[^{3}H]Methionine.

Because the f[^{3}H]methionine release assay is simpler and allows assay of rates of release and detection of both release factors (see below), only this assay is described in detail. Using this assay, Scolnick et al. [3] fractionated from E. coli B S-100 two release factors: Rl, which caused release upon translation of UAA or UAG but not UGA, and R2, which caused release upon translation of UAA or UGA but not UAG. Rl corresponds to the release factor reported by Capecchi, which led to release at the R17 mRNA codon UAG [4].

II. PREPARATION OF MATERIALS

A. Escherichia coli

Escherichia coli B has been used almost exclusively because its ribosomes can be used to bind f[^{3}H]Met-tRNA without initiation factors. Cells can be purchased from Grain Processing Company, Muscatine, Iowa, or grown to early log phase at 37°C with aeration in 1% dextrose and 0.8% nutrient broth.

B. Ribosomes

Escherichia coli B ribosomes are prepared by five to seven successive washes in NH_4Cl as described by Lucas-Lenard and

Lipmann [5] and are devoid of chain initiation, elongation, or termination factors. The first two washes should be in 1.0 M NH_4Cl, the remainder in 0.5 M NH_4Cl. An appropriate buffer is 0.01 M tris-HCl (pH 7.8)-0.001 M magnesium acetate-0.006 M 2-mercaptoethanol. Each time, the ribosomal pellet derived from a 100,000 X g ultracentrifugation at 4°C is resuspended in the above buffer and resedimented at 100,000 X g for 5-7 hr. The supernatant fluid is discarded. After the final wash ribosomes are resuspended in a buffer containing 0.01 M tris-HCl (pH 7.8), 0.05 M NH_4Cl, 0.01 M $MgCl_2$, and 0.006 M 2-mercaptoethanol at a concentration of 500-700 A_{260} units/ml. They can be stored indefinitely at -170°C with no loss of activity.

C. Codons

Trinucleotide codons AUG, UAA, UAG, and UGA may be purchased from Miles Research Company, or prepared by previously described methods [6].

D. f[^{3}H]Met-tRNA^f

Unfractionated E. coli tRNA can be purchased from Schwarz Bioresearch. It is best to purify partially the Met-tRNA^f species from it for use in the assay. This can be done by following the benzoylated DEAE-cellulose procedure of Gillam et al. [7] (see Chapter 10) using benzoylated DEAE-cellulose also purchased from Schwarz Bioresearch. f[^{3}H]Met-tRNA is prepared from this tRNA by using a preparation of synthetase

8. Preparation of Polypeptide Termination Factors

and transformylase prepared by the procedure of Muench and Berg [8] (see Chapter 9). Each reaction is incubated at 30°C for 30 min and contains in 1.0 ml: Leucovorin (Lederle Company), 0.24 mg; tRNA^{fMet}, 5-30 A_{260} units; enzyme protein, 0.5-1.0 mg; 0.1 M potassium cacodylate (pH 6.9); 1 X 10^{-3} M ATP; 0.01 M $MgCl_2$; 0.006 M 2-mercaptoethanol; and 5 X 10^{-5}-1.0 X 10^{-4} M [^{3}H]methionine and 19 other nonradioactive amino acids. The f[^{3}H]Met-tRNA is deproteinized, isolated, and stored as previously described [9].

The acylation reaction is terminated by the addition of 0.1 vol of 20% (w/v) potassium acetate (pH 5.5) and 1.1 vol of aqueous saturated phenol (Fischer, grade A). The solution is mixed on a vortex for 3-5 min and then the phases are separated by spinning at room temperature at 2000 rpm. The upper (aqueous) phase is aspirated and reextracted twice more with phenol as above. The three combined lower (phenol) phases are then back-extracted with 0.1 ml of 20% potassium acetate and 1.0 ml of water. The combined aqueous phases are then desalted on a G-25 (medium) Sephadex column equilibrated in 1 X 10^{-4} M potassium cacodylate (pH 5.5). The application volume should be no more than one-fifth the void volume of the column and fractions equal to the application volume can be collected. The tRNA is recovered in the void volume of the column and can be monitored by counting 0.010 ml of each fraction in Biosolv scintillation fluid (see below). The phenol and other salts are retarded by the column and are 5 to 10 fractions removed from the tRNA by this procedure. The acylated tRNA is stored lyophilized in convenient aliquots.

III. f[^{3}H]MET-tRNA-AUG-RIBOSOME COMPLEX

The precise level of ribosomes and fMet-tRNA used for formation of the intermediate varies with the particular ribosome preparation. Preliminary studies should be made to optimize binding of fMet-tRNA to the ribosome preparation. This may be done as follows. A reaction in 0.1 ml can be prepared with: AUG, 0.05 A_{260} units; 0.01 M magnesium acetate; 0.05 M tris-acetate (pH 7.2); 0.05 M ammonium acetate; f[^{3}H]Met-tRNA, 100 pmoles; and various levels of ribosomes, 2-30 A_{260} units. Reactions are incubated at 24°C for 15 min and the extent of binding determined by the assay of Nirenberg and Leder [10]. Care should be taken to plate on a Millipore filter no more than 2.0 A_{260} units of ribosomes so as not to exceed the capacity of the filter. Therefore, at high ribosome levels, aliquots of the original reaction (0.010 ml) must be taken. When the level of ribosomes is found at which 90% of the input tRNA is bound, these conditions can be used to prepare larger amounts of f[^{3}H]Met-tRNA-AUG-ribosome complex. If a reaction of 0.10 ml is prepared, at the end of 15 min at 24°C it is placed on ice and 0.40 ml of cold buffer is added that contains 0.05 M tris-acetate (pH 7.2), 0.05 M ammonium acetate,

8. Preparation of Polypeptide Termination Factors

and 0.035 M magnesium acetate. The high concentration of magnesium is needed to stabilize the complex as it is diluted from 0.10 to 0.50 ml. Aliquots of 0.015 ml are examined for the quantity of $f[^3H]$Met-tRNA-AUG-ribosome intermediate by the assay of Nirenberg and Leder; $f[^3H]$Met-tRNA by precipitation with cold trichloroacetic acid (TCA); and $f[^3H]$methionine as described below. Typical values are 4-6 pmoles of $f[^3H]$Met-tRNA-AUG-ribosome intermediate; 5-7 pmoles of $f[^3H]$Met-tRNA; and 0.15-0.40 pmole of $f[^3H]$methionine. The complex can be stored indefinitely at -170°C and the values do not change with storage for months with only one to three freeze-thaw cycles. This quantity of complex is adequate for 30 to 50 release assays.

IV. RELEASE ASSAY

Each 0.05 ml release assay mixture contains: $f[^3H]$Met-tRNA-AUG-ribosome complex, 4-6 pmoles; 0.05 M tris-acetate (pH 7.2); 0.03 M magnesium acetate; 0.10 M ammonium acetate; terminator codon, 0.01-0.10 A_{260} units; and a source of release factor. The mixture is incubated at 24°C for 15-30 min. To stop the reaction 0.250 ml of 0.10 M HCl is added; then 1.5 ml of ethyl acetate are added and the reaction is agitated for 10 sec on a vortex mixer; 1.0 ml of the ethyl acetate (top phase) is removed and transferred to 10.0 ml of BioSolv counting fluid (3 liters of toluene, 100 ml of Beckman BioSolv,

300 ml of Beckman Fluroalloy) and counted in a liquid scintillation counter. [^{3}H]Methionine released from f[^{3}H]Met-tRNA-AUG-ribosome complex is extracted into ethyl acetate at 70% efficiency under these conditions; f[^{3}H]Met-tRNA^f, which remains polar at pH 1.0, is not extracted. Values of nonspecific deacylation during the reaction can be determined by running parallel reactions without terminator codons.

V. PREPARATION OF R1 AND R2

By the use of this assay, R1 and R2 can be obtained from the S-100 fraction of E. coli extracts. They are present in low levels (about 500 molecules per cell) and therefore relatively large amounts of E. coli are required. All steps are performed at 4°C. One pound of E. coli cells is suspended in 681 ml of buffer A [0.10 M tris-acetate (pH 7.8), 0.05 M NH$_4$Cl, 0.014 M MgCl$_2$, 1 X 10^{-3} M dithiothreitol (DTT)]. The presence of the reducing agent DTT (or 2-mercaptoethanol or glutathione) is critical since release factors contain free sulfhydryl groups essential for their activity and the reducing agents protect them from air oxidation. The cells are lysed at 18,000 psi in a French pressure cell, and 3.0 mg/ml Worthington diisopropyl fluorophosphate-treated DNase are added. The diisopropyl fluorophosphate inactivates any proteolytic enzymes contaminating the DNase. The extract is centrifuged at 30,000 X g for 20 min, and the upper two-thirds to three-fourths

8. Preparation of Polypeptide Termination Factors

separated and centrifuged 5 hr at 137,000 X g. The upper two-thirds to three-fourths supernatant from this centrifugation corresponds to fraction I. A reliable specific activity for R is difficult to obtain with this fraction, but R activity can be measured on 5-30 μg of protein.

Both R1 and R2 are precipitated from fraction I by the addition of 32.5 g $(NH_4)_2SO_4$ per 100 ml. The precipitated R (fraction II, 28.0 g of protein) is collected by centrifugation (15 min at 30,000 X g), dissolved in 75 ml of buffer B [0.05 M tris-HCl (pH 8.0), 0.20 M KCl, 10^{-3} M DTT, 10^{-3} M EDTA], and applied to a 5 X 100 cm Sephadex G-100 (bead form) column equilibrated with buffer B. Each 12-ml fraction is eluted at a flow rate of 70 ml/hr and R activity is found in tubes 40 to 65. R1 and R2 do not separate by this procedure and tubes may be assayed with only UAA as messenger (the common release codon for both R factors). The active tubes are pooled (fraction III, about 2.0 g of protein) and applied to a 5 X 67 cm column packed with DEAE-Sephadex A-50, that has been freed of fine particles, equilibrated in buffer B. The protein is eluted, at whatever maximum flow rate is attainable, in 4200 ml of buffer B containing a linear KCl gradient from 0.15 to 0.70 M salt. An aliquot of 0.010 ml of the 15.0-ml fractions is adequate for determining release activity in a 30-min assay. R1 elutes at approximately 0.30 M KCl and R2 at 0.38 M KCl. R1 (IV) and R2 (V) are now fully separated and show their characteristic activity, R1 with UAA or UAG, R2 with UAA or UGA.

Release factor 1 (fraction IV) can be further purified by the following procedure. R1 (about 400 mg of protein) is equilibrated by dialysis in 0.05 M imidazole (pH 6.0)-0.01 M KCl-1 X 10^{-3} M EDTA (buffer C) and applied to a 40 X 2.5 cm column packed with carboxymethyl-Sephadex equilibrated in buffer C. The column is washed with 200 ml of buffer C and R1 is eluted with 600 ml of buffer containing a linear KCl gradient of 0.10-0.80 M salt, at a flow rate of 45 ml/hr. Release activity of each 7.5-ml fraction is assayed (0.010-ml aliquot) with UAA or UAG. The active fractions are pooled and R1 is dialyzed against 0.02 M tris-HCl (pH 8.0)-0.05 M KCl-0.001 M DTT. It is stored at -170°C (fraction VI) with no appreciable loss of activity with one to three freeze-thaw cycles.

Release factor 2 can be further purified as follows. R2 is equilibrated by dialysis in 0.01 M KH_2PO_4-K_2HPO_4 (pH 7.2)-1 X 10^{-3} M DTT and applied to a 9 X 2 cm column of hydroxylapatite (Clarkson Chemical Company) equilibrated in the same buffer (buffer D). From 16 to 500 mg of protein have been applied to this size column. The column is washed with 60 ml of, successively, 0.01 M, 0.035 M, and 0.08 M KH_2PO_4-K_2HPO_4 buffer (pH 7.2) with 1 X 10^{-3} M DTT. Fractions of 5.0 ml are collected. Since phosphate inhibits the release assay, either fractions should be dialyzed to remove phosphate before assay, or aliquots no greater than 0.002 ml should be directly assayed for release

8. Preparation of Polypeptide Termination Factors

activity. R2 elutes in the 0.08 M KH_2PO_4-K_2HPO_4 fraction (fraction VII).

All release factor preparations (IV through VII) are concentrated by pressure filtration (Amicon Company, Lexington, Massachusetts), dialyzed against 0.02 M tris-HCl (pH 8.0)-0.05 M KCl-3 X 10^{-3} M DTT, and stored at -170°C. They are stable at -170°C for at least a year and tolerate one to three freeze-thaw cycles well. Fractions VI and VII are each more than 1000-fold purified and are devoid of tRNA and other factors needed in initiation or elongation of protein synthesis. R2 is about 50% pure by acrylamide gel analysis; R1 is about 20% pure.

A. Other Assays

Although not described here, fractions VI (R1) and VII (R2) are suitable for assay of their activity by alternate means which have been used to dissect the mechanism of termination [11-14]. In summary, protein factors R1 and R2 bind to terminator codons on ribosomes and then, by a mechanism not yet elucidated, lead to the hydrolysis of the growing peptide chain from its peptidyl-tRNA linkage.

REFERENCES

[1] M. R. Capecchi, Proc. Natl. Acad. Sci. U.S., 58, 1144 (1967).
[2] C. T. Caskey, R. Tompkins, E. Scolnick, T. Caryk, and M. Nirenberg, Science, 162, 135 (1968).

[3] E. M. Scolnick, R. Tompkins, C. T. Caskey, and M. Nirenberg, Proc. Natl. Acad. Sci. U.S., 61, 768 (1968).

[4] A. Beaudet and C. T. Caskey, Nature, 227, 38 (1970).

[5] J. Lucas-Lenard and F. Lipmann, Proc. Natl. Acad. Sci. U.S., 57, 1050 (1967).

[6] P. Leder, M. Singer, and R. Brimacombe, Biochemistry, 4, 1561 (1967).

[7] T. Gillam, S. Millward, D. Blew, M. von Tigerstrom, E. Wimmer, and G. M. Tener, Biochemistry, 6, 3043 (1967).

[8] K. H. Muench and P. Berg, in Procedures in Nucleic Acid Research, Harper, New York, 1966, p. 375.

[9] C. T. Caskey, A. Beaudet, and M. Nirenberg, J. Mol. Biol., 37, 99 (1964).

[10] M. Nirenberg and P. Leder, Science, 145, 1399 (1964).

[11] E. Scolnick and C. T. Caskey, Proc. Natl. Acad. Sci. U.S., 64, 1235 (1969).

[12] R. Tompkins, E. Scolnick, and C. T. Caskey, Proc. Natl. Acad. Sci. U.S., 65, 702 (1970).

[13] J. Goldstein, G. Milmam, E. Scolnick, and C. T. Caskey, Proc. Natl. Acad. Sci. U.S., 65, 430 (1970).

[14] M. R. Capecchi and H. A. Klein, Cold Spring Harbor Symp. Quant. Biol., 34, 469 (1969).

Chapter 9

PREPARATION OF AMINOACYL-tRNA SYNTHETASES FROM Escherichia coli

Karl H. Muench

Division of Genetic Medicine
Departments of Medicine and Biochemistry
University of Miami School of Medicine
Miami, Florida

I. INTRODUCTION 214

II. METHODS OF ASSAY 216

    A. Amino Acid-Dependent ATP-PP_i Exchange. 217

    B. Formation of Aminoacyl-tRNA. 218

    C. Formation of Hydroxamates. 220

III. GENERAL APPROACH TO PURIFICATION OF SYNTHETASES. . . 221

    A. Use of Protective Agents 221

    B. Source of E. coli. 222

    C. Cell Disintegration and Centrifugation 223

    D. Removal of Nucleic Acids 224

    E. Purification by Solubility or Adsorption 226

    F. DEAE-Cellulose Chromatography. 226

    G. Hydroxylapatite Chromatography 227

    H. Storage. 229

    REFERENCES 230

Copyright © 1971 by Marcel Dekker, Inc. No part of this work may be reproduced or utilized in any form or by any means, electronic or mechanical, including xerography, photocopying, microfilm, and recording, or by any information storage and retrieval system, without the written permission of the publisher.

I. INTRODUCTION

Aminoacyl-tRNA synthetases or L-amino acid:tRNA ligases (AMP) catalyze reversible amino acid activation:

L-Amino acid + ATP + enzyme $\rightleftarrows$ Aminoacyl-AMP-enzyme + PP_i (1)

to form an enzyme-bound anhydride intermediate. The aminoacyl residue is then transferred to tRNA:

Aminoacyl-AMP-enzyme + tRNA $\rightleftarrows$ Aminoacyl-tRNA + AMP + enzyme.
(2)

Each enzyme and tRNA is specific for one L-amino acid.

Several excellent and comprehensive reviews describe aminoacyl-tRNA synthetases from all sources [1-5]. In this discussion we focus on enzyme preparations from E. coli. In general, the methods to be described have been successfully applied to purification of aminoacyl-tRNA synthetases from other bacteria, yeast, higher plants, mammals, and man.

Procaryotic organisms have a single synthetase for each amino acid. The number of Tyr-tRNA synthetase molecules in a single E. coli cell is approximately 1000, the same as the number of $tRNA^{Tyr}$ molecules [6]. In Salmonella typhimurium the concentration of His-tRNA synthetase is 2.2×10^{-6} M and of $tRNA^{His}$ 2.3×10^{-6} M, whereas the relevant K_M is 1.1×10^{-7} M [7]. This K_M value is typical and suggests that aminoacyl-tRNA synthetases may exist in vivo as complexes with their tRNAs [7]. The 20 aminoacyl-tRNA synthetases comprise an estimated 10% of the active enzyme mass of growing E. coli [8].

9. Aminoacyl-tRNA Synthetases from Escherichia coli

Table 1 shows the purifications so far reported for 19 aminoacyl-tRNA synthetases of E. coli and 1 of S. typhimurium; 12 have been purified to homogeneity. From these data a general approach to the purification of aminoacyl-tRNA synthetases can be derived.

Table 1

Purification of 20 Aminoacyl-tRNA Synthetases

Amino acid	Purification (fold)	Purity	Reference
Alanine	17	--	[9]
Arginine	500	--	[10]
Arginine	a	Homogeneous	[11]
Asparagine	40[b]	--	[12]
Aspartic acid	10[b]	--	[12]
Cysteine	40[b]	--	[12]
Glutamic acid	380	--	[13]
Glutamine	1000	Nearly homogeneous	[14]
Glycine	450	Homogeneous	[15]
Histidine[c]	800	Homogeneous	[7]
Isoleucine	300	Homogeneous	[16]
Leucine	560	95%	[17]
Leucine	850	Homogeneous	[18]
Lysine	850	Homogeneous	[19]
Lysine	800	Homogeneous	[20]

Table 1 (continued)

Amino acid	Purification (fold)	Purity	Reference
Methionine	290	Homogeneous	[21]
Methionine	780	Homogeneous	[22]
Methionine	675	Homogeneous	[23]
Phenylalanine	160	93%	[24]
Phenylalanine	350	Homogeneous	[25]
Proline	250	--	[26]
Serine	320	Homogeneous	[27]
Threonine	320	--	[28]
Tryptophan	1000	Homogeneous	[29]
Tyrosine	600	Homogeneous	[6]
Valine	640	Homogeneous	[30]

[a] Value not given.
[b] Value is approximate.
[c] This purification was from S. typhimurium.

II. METHODS OF ASSAY

Three different assays have been used to measure the activity of these enzymes. One measures only amino acid activation [Reaction (1)], whereas the other two measure the subsequent transfer of the activated aminoacyl residue to an acceptor. In one assay this acceptor is hydroxylamine; in the other assay the acceptor is a specific tRNA [Reaction (2)].

9. Aminoacyl-tRNA Synthetases from Escherichia coli

A. Amino Acid-Dependent ATP-PP$_i$ Exchange

The rate of incorporation of [^{32}P]PP$_i$ into ATP by the reverse of Reaction (1) is conveniently measured at 23-37°C in a 1.0-ml reaction mixture containing 2 mM ATP; 2 mM [^{32}P]PP$_i$, with a specific activity of 10^4-10^5 cpm/µmole; 5 mM MgCl$_2$; 100 mM tris-HCl or potassium bicine buffer (pH 7.5-8.5); 0 or 2 mM L-amino acid; and enzyme. Under these conditions the rate of the reaction is essentially linear until 0.6 µmole of [^{32}P]PP$_i$ has been incorporated into ATP [31]. The reaction is stopped with 0.5 ml of 7% perchloric acid, followed by 0.2 ml of a 15% (w/v) suspension of acid-washed amorphous carbon (Norite). The Norite is collected by filtration on a glass-fiber filter (Whatman, GF/C) and washed five times, each with 5 ml of water, in an appropriate filter holder attached to a water suction pump. The moist filter can be directly glued (DuPont Duco Cement) to a planchet, dried, and counted in a gas-flow Geiger counter, or the filter can be dried, immersed in liquid scintillator, and counted in a scintillation counter (see Chapter 6).

In order to inhibit inorganic pyrophosphatase present in the early stages of purification, 10 mM KF is added to the reaction mixture [32,33]. Ordinarily, KF does not inhibit ATP-PP$_i$ exchange. The control measurement of ATP-PP$_i$ exchange that is not dependent on the added L-amino acid is particularly important in cell extracts and cruder fractions of the enzyme purification procedure.

There are several advantages to this method of assay. $[^{32}P]PP_i$ is inexpensive either when purchased directly or prepared [34,35]. Because $[^{32}P]PP_i$ can be made with very high specific activity, and because the ratio of amino acid activation to amino acid transfer is of the order 10 to 100, the assay can be extremely sensitive. Moreover, the assay can be performed with little or no net change in the concentrations of substrates in the assay medium. This assay circumvents the need for tRNA, which may be difficult to obtain in large amounts from the same source as the enzyme under study. Moreover, because a given enzyme may lose its aminoacyl-tRNA synthetic activity without losing its ability to catalyze amino acid-dependent ATP-PP_i exchange, only this assay detects the altered enzyme. In this way a particular conformation of Leu-tRNA synthetase unable to catalyze Leu-tRNA formation was discovered and described [18]. Finally, in terms of instrumentation, this method easily lends itself to measurement by a gas-flow Geiger counter (which is less expensive to buy and to operate than a liquid scintillation counter).

B. Formation of Aminoacyl-tRNA

This assay is conveniently done at 23-37°C in a 0.5-ml reaction mixture containing 100 mM buffer, 2-20 mM $MgCl_2$, 1 mM ATP, 0.1 mM L$[^{14}C]$amino acid or L$[^{3}H]$amino acid containing 10^7 cpm/μmole, 20 A_{260} units of unfractionated tRNA containing

9. Aminoacyl-tRNA Synthetases from <u>Escherichia coli</u>

50 pmoles of specific acceptor per A_{260} unit, and enzyme. The concentration of specific tRNA is then 2 µM. The buffer may be sodium or potassium cacodylate in the pH range 6.5-7.0, tris cacodylate in the pH range 6.5-8.5, tris-HCl in the pH range 7.5-8.5, or sodium or potassium bicine in the pH range 7.8-8.8. In general, pH optima for aminoacyl-tRNA synthetases are higher than pH 7.5, but at higher pH values aminoacyl-tRNAs are unstable and the rate of chemical hydrolysis must be considered. The reaction is stopped with 3 ml of 2 M HCl after 2-10 min. The acid-insoluble material is collected on glass-fiber filters (Whatman GF/C), washed five times, each with 3 ml of 2 M HCl, dried, immersed in scintillation fluid, and counted. The formation of product is linear up to about 0.5 nmole.

The aromatic amino acids tyrosine, phenylalanine, and tryptophan tend to give high blanks in assay tubes lacking enzyme. The high blanks may be lowered by preliminary purification of the precursor radioactive amino acids and by the addition of an alcohol precipitation step after the assay. Thus L[^{14}C]tryptophan is first purified by chromatography on Dowex-1 [36]. The reaction is stopped by the addition of 2.6 vol of a solution made from 1 part of 2 M potassium acetate buffer (pH 5) and 12 parts of ethanol. The resulting precipitate is collected by centrifugation, dissolved in 0.1 M potassium acetate buffer (pH 5), and then subjected to the HCl precipitation and collection on glass-fiber filters already described.

The assay based on formation of aminoacyl-tRNA has several advantages. Of the three assays described, it is closest to the action of the synthetases in vivo. The reaction is more specific than amino acid activation. For example, Ile-tRNA synthetase of *E. coli* activates both L-isoleucine and L-valine, and Val-tRNA synthetase of *E. coli* activates both L-valine and L-threonine, but these enzymes place only the correct amino acid into ester linkage on tRNA [33]. The assay of aminoacyl-tRNA synthesis is better suited for kinetic studies under the Michaelis-Menten assumptions because in the ATP-PP$_i$ exchange reaction the various substrates are in equilibrium and the enzyme exists as a complex with the aminoacyl-AMP product [37]. In some cases [38-41] the presence of tRNA is required for amino acid activation. In terms of instrumentation the formation of aminoacyl-tRNA is better adapted to scintillation counting than is the ATP-PP$_i$ exchange assay.

C. Formation of Hydroxamates

Hydroxylamine serves as an acceptor of activated amino acids in place of tRNA; the resulting hydroxamates form colored complexes with ferric ion. The reaction is important historically [32], but it is insensitive and requires concentrations of hydroxylamine in the range 1-4 M. The sensitivity and specificity is improved by the use of radioactive amino acids with subsequent separation of the radioactive hydroxamates on

9. Aminoacyl-tRNA Synthetases from Escherichia coli 221

ion-exchange paper [42]. Although Thr-tRNA synthetase of E. coli does not make threonine hydroxamate, the compound is made via a threonyl-AMP-E^{Val} complex and via Thr-tRNA [43].

III. A GENERAL APPROACH TO PURIFICATION OF SYNTHETASES

A. Use of Protective Agents

Various agents, when present in the buffer solutions used for purification of the enzymes, increase the recovery of activity. In most cases the inclusion of a reducing agent such as 2-mercaptoethanol, glutathione, or dithiothreitol (DTT) is desirable. Of the 20 aminoacyl-tRNA synthetases in E. coli, only Lys-tRNA synthetase is impervious to the action of agents reacting with sulfhydryl groups [44]. The use of 20-50 mM 2-mercaptoethanol is common and provides increased recoveries of many aminoacyl-tRNA synthetases. However, 2-mercaptoethanol should not be used with cacodylate buffer because they react with a consequent alteration of the pH [45]. When DTT is substituted for 2-mercaptoethanol, the recovery of Thr-tRNA synthetase is doubled [28].

The presence of 10% glycerol in buffers protects at least six aminoacyl-tRNA synthetases: those for glycine, leucine, phenylalanine, proline, serine, and threonine; in no case so far reported does glycerol harm any of the aminoacyl-tRNA synthetases [9,17,24,26-28]. Pro-tRNA synthetase is a special case because its relatively unstable subunits form a stable dimer in the presence of glycerol [26]. Ser-tRNA synthetase may do the same [27]. Propylene glycol is less viscous and may be a satisfactory substitute for glycerol [46]. Sucrose stabilizes Pro-tRNA synthetase but must be used at higher concentrations than glycerol [26].

Although EDTA is frequently used in chromatographic buffers [24], it can harm at least one activity, Leu-tRNA synthetase [17]. EDTA cannot be used in chromatography on hydroxylapatite columns. The Mg^{2+} ion protects at least one activity; in the absence of Mg^{2+} Leu-tRNA synthetase is unstable.

Although several aminoacyl-tRNA synthetases are stabilized by substrates, they have not been widely used during purification. Arginine at 0.1 mM was used in the purification of two mutant Arg-tRNA synthetases [47].

B. Source of E. coli

We have used E. coli strains B and K12, grown in minimal medium [9], harvested in log growth, and stored at -20°C, in which little or no loss of activity occurs over long periods. Commercial cells, supplied washed or unwashed as a frozen paste (Grain Processing Company, Muscatine, Iowa) are an excellent source of enzymes [12].

In the subsequent sections the paragraphs entitled Procedure form a sequence of specific instructions for a purification applicable to all 20 aminoacyl-tRNA synthetases of

9. Aminoacyl-tRNA Synthetases from Escherichia coli

E. coli. These paragraphs are followed by relevant discussions, and alternative and additional purification procedures are indicated where appropriate. Except as noted, all purification steps are at 0-2°C.

C. Cell Disintegration and Centrifugation

Procedure. Escherichia coli paste is suspended in 1-2 vol of 0.01 M tris-HCl buffer (pH 8.0)-0.01 M $MgCl_2$-10% glycerol. The suspension is passed through a French press at 8000 psi. The lysed suspension is centrifuged for 1 hr at 198,000 X g and the entire supernatant solution (extract) harvested.

For the breakage of E. coli cells, several methods, for example, sonication, blending with glass beads, and passage through a French press, are successful. The methods of breakage, when optimized, all provide approximately the same total activity in crude extracts of E. coli. Therefore the difficulty is not great in scaling up from a method best suited to small quantities of extract to a method capable of providing enzyme sufficient for chemical studies. The advantage of the French press [48] (American Instrument Company) is that each cell in suspension is subjected only once to breaking force. In most other methods cells randomly encounter the breaking force. To insure 99% breakage the force must be applied for a relatively long time, and enzymes from cells broken early are subjected to denaturation. Therefore a time must be found at which the difference, enzyme release minus enzyme denaturation, is maximal.

The use of an ultracentrifuge to remove insoluble material is ideal for small preparations but does not lend itself to large preparations. For larger preparations conventional centrifugation, for example, 2 hr at 16,000 X g, is satisfactory. Clarification of the then-turbid supernatant liquid occurs during the subsequent removal of nucleic acids as described in Section II,D.

D. Removal of Nucleic Acids

Procedure. The extract is passed over a DEAE-cellulose column equilibrated with 10 mM KH_2PO_4-K_2HPO_4 buffer (pH 6.9) in solution A (20 mM 2-mercaptoethanol, 1 mM $MgCl_2$, 10% glycerol). The DEAE-cellulose (0.9 meq/gm), when packed at 5 psi, should occupy a volume in milliliters at least 7 times the weight in grams of the E. coli paste used to prepare the extract. To insure adsorption of Trp-tRNA synthetase, the applied extract should have a conductivity (essentially, the reciprocal of salt concentration) of less than 1 mmho. The column is washed with 5 vol of the equilibration buffer and the aminoacyl-tRNA synthetases eluted with 250 mM KH_2PO_4 buffer (pH 6.9) in solution A. The peak of A_{280} resulting from the change of buffer contains the aminoacyl-tRNA synthetases in one-third to two-thirds of a column volume (DEAE-cellulose batch fraction).

9. Aminoacyl-tRNA Synthetases from Escherichia coli

For sharp fractionation by subsequent techniques, the extract must first be freed of nucleic acids. At least six different approaches are available, and some of these may provide enzyme purification in terms of protein as well. The method described is applicable to all aminoacyl-tRNA synthetases and to ATP, CTP:tRNA nucleotidyl transferase as well. The method removes tRNA, amino acids, and almost all nuclease activity, which does not adhere to the column. Some ribonuclease II may not be removed, however [49].

Other methods, which may offer advantages for a particular enzyme, include precipitation of nucleic acids with streptomycin sulfate [10,19,20,23,24,30], protamine sulfate [13], or $MnCl_2$ [29,33]. Alternatively, nucleic acids can be removed by differential partitioning between aqueous polymer solutions [17].

For stable enzymes incubation of the extract for 3-5 hr at 37°C in the presence of phosphate buffer leads to the breakdown of nucleic acids [6,16,22,27,28]. Pro-tRNA synthetase, which is converted into a stable dimer at 37°C [26], seems to be a natural choice for this "autolysis." In practice, however, the procedure is not suitable for Pro-tRNA synthetase purification since, although there is no loss of activity during the autolysis, in the later stages of purification the contaminating proteins are more difficult to remove. Autolysis may change the quaternary structure of aminoacyl-tRNA synthetases composed of subunits. Met-tRNA synthetase has a molecular weight of approximately 180,000 [21,23] when purified without an autolysis step, and a molecular weight of 96,000 when purified [22] with an autolysis step; autolysis breaks the enzyme into active subunits [23].

E. Purification by Solubility or Adsorption

Prior to column chromatography it may be desirable to obtain an initial purification by means of differential precipitation with $(NH_4)_2SO_4$ [6,10,16,20,23,24,28,30], or by differential adsorption to calcium phosphate gel [10,13,19,25,30] or alumina C-γ gel [6,16,28,33].

F. DEAE-Cellulose Chromatography

Procedure. The DEAE-cellulose batch fraction, dialyzed to equilibrium with 10 mM KH_2PO_4-K_2HPO_4 buffer (pH 6.9) in solution A, is applied to a DEAE-cellulose (Whatman DE-52) column packed by gravity alone and equilibrated with the same buffered solution. At least 7 mg of protein in this fraction may be applied for every 1-ml column volume of DEAE-cellulose. The column is washed with 1 column volume of the equilibrating buffer and developed with a linear gradient from 20 to 250 mM KH_2PO_4-K_2HPO_4 buffer (pH 6.9) in 30 column volumes of solution A. The desired peak of enzyme activity is located by assay, and the appropriate fractions are pooled and concentrated by dialysis against the initial equilibration buffer containing 5-30% polyethylene glycol (molecular weight 6000) (DEAE-cellulose gradient fraction).

The A_{280} pattern in the effluent is reproducible and allows easy location of specific enzyme peaks [12]. The elution sequence is : ATP, CTP:tRNA nucleotidyl transferase, then the aminoacyl-tRNA synthetases for tryptophan, alanine; aspartate, glutamate; isoleucine; tyrosine; cysteine, methionine, phenylalanine; glycine, proline, threonine, arginine; histidine, lysine; serine; asparagine; valine; and leucine. Enzymes between semicolons are not well separated from others in the same group. The position of Gln-tRNA synthetase in this sequence is not known. Asparaginase that contaminates the Asn-tRNA synthetase purified to this stage can be removed by subsequent chromatography on DE-52 at pH 8.1 [12]. In many cases sequential DEAE-cellulose columns developed with gradients at different pH values have led to additional purifications [6,12,17,23].

G. Hydroxylapatite Chromatography

Procedure. The DEAE-cellulose gradient fraction is applied to a hydroxylapatite column equilibrated with 10 mM KH_2PO_4-K_2HPO_4 buffer (pH 6.9)-20 mM 2-mercaptoethanol-10% glycerol. The column is washed with several volumes of the buffer and then developed with a linear gradient from 10 to 200 mM KH_2PO_4-K_2HPO_4 buffer (pH 6.9) in 50 column volumes of 20 mM 2-mercaptoethanol-10% glycerol. The peak of enzyme activity is located by assay, and the appropriate fractions are pooled and concentrated by dialysis against solution A containing 10 mM KH_2PO_4-K_2HPO_4 buffer (pH 6.9) and 15-30% (w/v) polyethylene glycol (hydroxylapatite fraction).

The flow rates achievable on an hydroxylapatite column are dependent on its delicate crystal structure. Because of the mechanical fragility of hydroxylapatite crystals, commercial hydroxylapatite may be harmed during shipment, and this may account for the variability from batch to batch in commercial preparations. We use only hydroxylapatite prepared in our own laboratory as the "CPA" material described by Main et al. [50]. A convenient procedure for preparation of this hydroxylapatite has been given [51].

At least 4 mg of the DEAE-cellulose batch fraction may be applied for every 1 ml of hydroxylapatite column volume. The enzyme elution sequence [12], designated by the amino acid, is: proline, cysteine; serine; asparagine; aspartate, histidine, arginine; glycine; glutamine, valine, phenylalanine; tyrosine, methionine; glutamate, leucine; alanine, lysine; isoleucine; tryptophan; and threonine. The enzymes between semicolons are not well separated from the others in the same group. A similar but not identical elution sequence has been reported [43]. Lys-tRNA synthetase, for unknown reasons, may emerge as three peaks, one preceding and one following that designated [12]. The enzyme travels as two peaks on TEAE-cellulose chromatography [52]. As in the case of DEAE-cellulose,

9. Aminoacyl-tRNA Synthetases from Escherichia coli

hydroxylapatite columns used sequentially at two different pH values may give additional purification [17,23,24,27].

Sephadex G-200 [23,25], preparative polyacrylamide gel electrophoresis [27], and a wide variety of other ion-exchange resins, including ECTEOLA-cellulose [53], DEAE-Sephadex [7,20, 30], phosphocellulose [6,7], and polymethacrylate [20,29] have provided useful purifications of aminoacyl-tRNA synthetases. The procedure given here is emphasized because it extensively purifies all 20 aminoacyl-tRNA synthetases. For example, the method provides a 230-fold purification of Trp-tRNA synthetase with 44% recovery [54]. Most of the other methods are less general in their applicability.

H. Storage

Most of the aminoacyl-tRNA synthetases of E. coli are inactivated by repeated freezing and thawing in buffer solutions. Freezing in the presence of 20 mM 2-mercaptoethanol completely denatures these enzymes. The hydroxylapatite fraction can be stored at 0°C. Alternatively, the concentrated enzyme solution is brought to 50% in glycerol and stored at -20°C. The aminoacyl-tRNA synthetases we have examined are stable indefinitely under these conditions; for example, a preparation of Tyr-tRNA synthetase [6] so stored for 6 years has lost no activity.

REFERENCES

[1]   P. Berg, *Ann. Rev. Biochem.*, 30, 293 (1961).

[2]   G. D. Novelli, *Ann. Rev. Biochem.*, 36, 449 (1967).

[3]   P. J. Peterson, *Biol. Rev.*, 42, 552 (1967).

[4]   P. Lengyel and D. Söll, *Bacteriol. Rev.*, 33, 264 (1969).

[5]   R. B. Loftfield, in *Protein Synthesis*, Vol. 1 (E. McConkey, ed.), Marcel Dekker, in press.

[6]   R. Calendar and P. Berg, *Biochemistry*, 5, 1681 (1966).

[7]   F. DeLorenzo and B. N. Ames, *J. Biol. Chem.*, 245, 1710 (1970).

[8]   F. C. Neidhardt and L. S. Williams, *J. Mol. Biol.*, 43, 529 (1969).

[9]   K. H. Muench and P. Berg, in *Procedures in Nucleic Acid Research*, Vol. 1 (G. L. Cantoni and D. R. Davies, eds.), Harper, New York, 1966, p. 375.

[10] S. K. Mitra and A. H. Mehler, *J. Biol. Chem.*, 242, 5490 (1967).

[11] J. A. Haines and P. C. Zamecnik, *Biochim. Biophys. Acta*, 146, 227 (1967).

[12] K. H. Muench and P. A. Safille, *Biochemistry*, 7, 2799 (1968).

[13] R. A. Lazzarini and A. H. Mehler, *Biochemistry*, 3, 1445 (1964).

[14] W. R. Folk and P. Berg, *J. Bacteriol.*, 102, 193,204 (1970).

9. Aminoacyl-tRNA Synthetases from Escherichia coli

[15] D. L. Ostrem and P. Berg, Proc. Natl. Acad. Sci. U.S., 67, 1967 (1970).

[16] A. N. Baldwin and P. Berg, J. Biol. Chem., 241, 831 (1966).

[17] H. Hayashi, J. R. Knowles, J. R. Katze, J. Lapointe, and D. Söll, J. Biol. Chem., 245, 1401 (1970).

[18] P. Rouget and F. Chapeville, European J. Biochem., 14, 498 (1970).

[19] R. Stern and A. H. Mehler, Biochem. Z., 342, 400 (1965).

[20] J. Waldenstrom, European J. Biochem., 3, 483 (1968).

[21] R. L. Heinrikson and B. S. Hartley, Biochem. J., 105, 17 (1967).

[22] C. J. Bruton and B. S. Hartley, Biochem. J., 108, 281 (1968).

[23] F. Lemoine, J.- P. Waller, and R. van Rapenbusch, European J. Biochem., 4, 213 (1968).

[24] M. P. Stulberg, J. Biol. Chem., 242, 1060 (1967).

[25] M. H. J. E. Kosakowski and A. Böck, European J. Biochem., 12, 67 (1970).

[26] M. Lee and K. H. Muench, J. Biol. Chem., 244, 223 (1969).

[27] J. R. Katze and W. Konigsberg, J. Biol. Chem., 245, 923 (1970).

[28] D. I. Hirsh, J. Biol. Chem., 243, 5731 (1968).

[29] D. R. Joseph and K. H. Muench, Federation Proc., 29, Abst. 1246 (1970).

[30] M. Yaniv and F. Gros, J. Mol. Biol., 44, 1 (1969).

[31] P. Berg, J. Biol. Chem., 222, 1025 (1956).

[32] M. B. Hoagland, Biochim. Biophys. Acta, 16, 288 (1955).

[33] F. H. Bergmann, P. Berg, and M. Dieckmann, J. Biol. Chem., 236, 1735 (1961).

[34] C. H. L. Peng, Biochim. Biophys. Acta, 22, 42 (1956).

[35] A. Kornberg and W. E. Pricer, Jr., J. Biol. Chem., 191, 535 (1951).

[36] C. H. W. Hirs, S. Moore, and W. H. Stein, J. Am. Chem. Soc., 76, 6063 (1954).

[37] F. X. Cole and P. R. Schimmel, Biochemistry, 9, 480 (1970).

[38] J. M. Ravel, S. Wang, C. Heinemeyer, and W. Shive, J. Biol. Chem., 240, 432 (1965).

[39] L. W. Lee, J. M. Ravel, and W. Shive, Arch. Biochim. Biophys., 121, 614 (1967).

[40] A. H. Mehler and S. K. Mitra, J. Biol. Chem., 242, 5495 (1967).

[41] S. K. Mitra and C. J. Smith, Biochim. Biophys. Acta, 190, 222 (1969).

[42] R. B. Loftfield and E. A. Eigner, Biochim. Biophys. Acta, 72, 372 (1963).

[43] D. I. Hirsh and F. Lipmann, J. Biol. Chem., 243, 5724 (1968).

[44] R. Stern, M. DeLuca, A. H. Mehler, and W. D. McElroy, Biochemistry, 5, 126 (1966).

[45] J. R. Knowles, J. R. Katze, W. Konigsberg, and D. Söll, J. Biol. Chem., 245, 1407 (1970).

[46] F. J. Kull and K. B. Jacobson, Proc. Natl. Acad. Sci. U.S., 62, 1137 (1969).

[47] I. N. Hirshfield and H. P. J. Bloemers, J. Biol. Chem., 244, 2911 (1969).

[48] C. S. French and H. W. Milner, in Methods in Enzymology, Vol. 1 (S. P. Colowick and N. O. Kaplan, eds.), Academic Press, New York, 1955, p. 64.

[49] M. F. Singer and N. G. Nossal, J. Biol. Chem., 243, 913 (1968).

[50] R. K. Main, M. J. Wilkins, and L. J. Cole, J. Am. Chem. Soc., 81, 6490 (1959).

[51] K. H. Muench, in Procedures in Nucleic Acid Research, Vol. 2 (G. L. Cantoni and D. R. Davies, eds.), Harper, New York, 1971, in press.

[52] R. D. Marshall and P. C. Zamecnik, Biochim. Biophys. Acta, 181, 454 (1969).

[53] D. J. McCorquodale, Biochim. Biophys. Acta, 91, 541 (1964).

[54] K. H. Muench, Biochemistry, 8, 4872 (1969).

Chapter 10

TRANSFER RNA

Jack Goldstein

New York Blood Center
and
Cornell University Medical College
New York, New York

I. INTRODUCTION 235

   Nomenclature 237

II. ISOLATION OF CRUDE tRNA. 238

   A. From Whole Cells 238

   B. From Disrupted Cells 248

III. FURTHER PURIFICATION 250

IV. FRACTIONATION METHODS. 257

   A. Partition Systems. 257

   B. Miscellaneous Chromatographic Methods. 261

   REFERENCES 262

I. INTRODUCTION

Let us begin by briefly recounting the principal events that led to the discovery of tRNA.

Copyright © 1971 by Marcel Dekker, Inc. No part of this work may be reproduced or utilized in any form or by any means, electronic or mechanical, including xerography, photocopying, microfilm, and recording, or by any information storage and retrieval system, without the written permission of the publisher.

By the mid-1950s, initially as a result of the work of Hoagland [1], which was then amplified by several others [2-4] (see also Chapter 9), it was known that the first step in protein biosynthesis involved "activation" of amino acids, that is, the formation of an enzyme-bound mixed anhydride derivative of AMP and an amino acid [Equation (1)].

$$\text{Enzyme} + \text{amino acid} + \text{ATP} \rightleftharpoons \text{Enzyme-aminoacyl-AMP} + \text{pyrophosphate} \quad (1)$$

In 1956, Holley [5] suggested that as the next step the aminoacyl adenylate could react with some unknown acceptor (X) with the liberation of AMP, and postulated Equation (2).

$$\text{Enzyme-aminoacyl-AMP} + X \rightleftharpoons \text{Enzyme} + \text{aminoacyl-X} + \text{AMP} \quad (2)$$

Using an assay system that measured the incorporation of ^{14}C-labeled AMP into ATP [the reverse reactions of Equations (2) and (1)], he showed that X existed and, furthermore, that it was sensitive to pancreatic RNase. These results were confirmed and extended with purified components [6].

These observations fitted very well with the kinetic studies of Hultin and Beskow [7], which indicated the presence of an amino acid intermediate in protein biosynthesis, and the notion of Crick [8] that some sort of adapter molecule was necessary to act as an intermediary between the nucleic acid template and the amino acids being incorporated into the

10. Transfer RNA

growing polypeptide chain. Final proof of the identity of X was provided by Hoagland et al. [9], who isolated RNA charged with [^{14}C]leucine from a rat liver "pH 5 enzymes" preparation and showed that it could transfer the [^{14}C]leucine to microsomal protein. Because this material had been isolated from a nonparticulate cell fraction, the 105,000 X g supernatant, he referred to it as sRNA (soluble RNA).

Since then, of course, a great deal of work has been and is still being done with tRNAs. This research presently ranges from structure-function interrelationship studies to the possible role of tRNA as a control mechanism in the translation of genetic information (for details, see Ref. 10). It is the purpose of this chapter to acquaint the reader who may be interested in pursuing some of these studies with the techniques necessary for the isolation of tRNA and for its separation into individual amino acid acceptor species.

Nomenclature

The commonly used synonyms for transfer RNA or tRNA (soluble RNA or sRNA, referring to RNA nonsedimentable at 105,000 X g or soluble in 1 M NaCl; or 4 S RNA, RNA exhibiting a sedimentation coefficient of 4 S) are misleading since preparations so designated often contain RNAs other than tRNAs. Therefore we use the term crude tRNA instead. Also, the terms tRNAs and amino acid acceptor RNAs are used interchangeably.

It is also customary to refer to a specific nonacylated tRNA, for example, alanine tRNA or tRNA^{Ala} and its aminoacylated counterpart, as Ala-tRNA or Ala-tRNA^{Ala}. Isoacceptor species, defined as two or more tRNAs that accept the same amino acid, are designated by subscripts, for example, tRNA$^{Ala}_1$, tRNA$^{Ala}_2$, and so on. These rules essentially serve our purpose; for a more complete listing, the reader is referred to the recommendations of the IUPAC-IUB Commission on Biochemical Nomenclature [11].

II. ISOLATION OF CRUDE tRNA

A. From Whole Cells

Whole-cell preparations have the advantage of relative speed, simplicity, and high yields [12].

The method to be described for <u>Escherichia coli</u> is a modification of a procedure developed by Holley et al. [13] for large-scale preparations from yeast. It, in turn, was a modification of the original whole-cell method developed with yeast by Monier et al. [14]. These, and indeed most methods developed for the preparation of nucleic acid, employ phenol as the nucleic acid extractant and protein denaturant. The use of phenol dates back to a brief report on the isolation of RNA from tobacco mosaic virus by Gierer and Schramm [15] and to the more comprehensive paper by Kirby [16] on the isolation of RNAs from mammalian tissues, which really provided the technical basis for all succeeding work.

10. Transfer RNA

| Escherichia coli |

   Thaw at 4°C in water
   Add 1.2 vol of water-saturated phenol
   Extract 2 hr
   Centrifuge; discard phenol layer and interface material

| Aqueous Phase |

   Add 0.1 vol of 20% potassium acetate,
     followed by 2.5 vol of 95% ethanol
   Let stand 4 hr to overnight at -20°C
   Centrifuge and collect precipitate
   Partially dry under reduced pressure

| Precipitate |

   Suspend in 0.1 M tris-HCl (pH 7.5)
   Let stir 15 min
   Let stand 5 min; centrifuge (if necessary)
   Add suspension to DEAE-cellulose
   Let stir 4 hr
   Wash four times with 0.1 M tris-HCl (pH 7.5)
   Collect DEAE-cellulose

| DEAE-cellulose |

   Add 1 M NaCl in 0.25 M tris-HCl (pH 7.5)
   Let stir 4 hr
   Centrifuge and collect supernatant

| Supernatant |

   Add 2.5 vol of 95% ethanol
   Let stand 4 hr to overnight at -20°C
   Centrifuge and collect precipitate

| Precipitate |

   Wash twice with 80% ethanol;
     wash once with 95% ethanol
   Dry in vacuo

| Crude tRNA |

Diagram 1. DEAE-cellulose procedure for the preparation of crude tRNA from whole cells.

One can start (see Diagram 1) with either freshly prepared, frozen, or lyophilized E. coli cells grown to any stage (early, mid-, or late log) that have been washed free of growth media by centrifugation. Because of the presence of a "finger nuclease" [17], it is advisable that the operator wear gloves (the disposable, polyethylene variety or anything similar is quite suitable) throughout the whole procedure, and that all glassware and equipment involved be cleaned free of finger marks prior to use. A given weight of cells, wet or otherwise, is suspended in cold (4°C) deionized water (about 1 g of cells to 4-5 ml water). If frozen cells are used, they should be allowed to thaw in the suspending water in a cold room. The thawing can be somewhat hastened by intermittent stirring, either by hand with a glass rod or with a stirrer and magnetized Teflon stirring bar. In any case the stirring should be gentle, and suspension of the cells should be complete before the addition of phenol. Unsuspended clumps of cells tend to produce large clots of cells when phenol is added, which reduces the effectiveness of the extraction and consequently the yield. Phenol at cold-room temperature, previously saturated with deionized water, is slowly added (a total of 1.2 vol of water-saturated phenol per 1 vol of water is used to suspend cells) with very vigorous hand shaking or swirling (depending on volume involved); agitation is continued for 1-2 min after the final addition of the phenol. [There is no

10. Transfer RNA

need for redistillation if liquified phenol of reagent grade (preservative free) is used. It is convenient to purchase pint bottles and to discard unused material 3-4 weeks after opening. The use of phenol incompletely saturated with water in the extraction leads to a decrease in volume of the aqueous phase and loss of RNA.] This hand manipulation is necessary to insure complete contact between the suspended cells and droplets of phenol so that cell clumping is minimized and, more importantly, so that any nucleases released are immediately fully exposed to the denaturing effects of the phenol. Extraction is then allowed to continue for 2 hr in a cold room with a moderate amount of shaking or stirring. A "wrist-action" shaker of the type used for growing microorganisms is recommended for most extractions. When this is used, the extraction is usually carried out in Erlenmeyer flasks of 4 times the volume of the extraction mixture which are tightly covered with either rubber stoppers or corks wrapped in Parafilm or Saran Wrap. Very small extractions (1-10 ml of extraction mixture) may be effectively accomplished with a magnetized Teflon stirring "ball" and stirrer in an appropriately sized Kimax screw-top vial; if a screw cap is used for closure, it should be lined with Teflon or some other inert material.

At the end of the extraction period, the phases are separated by centrifuging the mixture at 12,000 X g for 20 min in a refrigerated centrifuge. When very large volumes, such as

4 liters or more, are involved and centrifugation is inconvenient, phases can be separated by allowing the mixture to stand overnight in a cold room. The aqueous phase (upper layer in this case) is carefully collected by gentle aspiration and its volume is measured (it is best not to collect further down than about 0.5 cm above the interface). The remainder of the mixture, including the last of the upper phase, all of the interface material, and the lower phase, is discarded. To the measured aqueous phase, in an ice bath or cold room, is now added, with mixing, 0.1 of its volume of cold 20% potassium acetate; this is then followed by the slow addition of 2.5 times its volume of cold 95% ethanol. (Ethanol should be reagent grade. Some of the phenol dissolved in the aqueous phase is removed by the ethanol; the presence of the remainder during the next stage of the isolation is not harmful and in fact may be beneficial in helping to protect against any traces of nucleases that may be accidentally introduced. Excess phenol can always be removed by extracting the aqueous phase three to four times with an equal volume of diethyl ether. A stream of nitrogen gas is then bubbled through the solution to remove any traces of ether.) The precipitation is completed by storage for at least 4 hr (preferably overnight) at -20°C (the ethanol precipitation steps are good overnight, or longer, stopping points in the preparative procedure) or, for very large work-ups, a minimum of 24 hr in a cold room.

10. Transfer RNA

The flocculent precipitate is collected by centrifugation at 12,000 X g for 20 min in a refrigerated centrifuge (prior to this, for large work-ups, some of the supernatant may be removed by decantation); the supernatant is removed and the precipitate is allowed to drain for about 10 min in a cold room by inverting the centrifuge tube. Any liquid adhering to the inside wall of the tube above the precipitate is removed by swabbing with a lint-free tissue. Residual ethanol is then removed from the precipitate until it is converted from a wet paste to a moist powder; total drying is to be avoided. This partial drying of the precipitate in the centrifuge tube is continued for a short time at room temperature (30 min to 2 hr, depending on the size of precipitate) in a vacuum desiccator connected to a water aspirator and that contains silica gel as the drying agent. The precipitate is a very heterogeneous mixture containing, in addition to tRNA as its major component, varying amounts of all the other cellular RNAs and RNA fragments, some polysaccharides, and very small amounts, if any, of fragmented DNA. Also present are any proteins, protein fragments, and peptides left sufficiently undenatured by the phenol treatment to be extracted into the aqueous phase. Complete drying, therefore, might promote undesirable interactions among the various components, so only enough ethanol is removed to insure that the solubility of the tRNA is not reduced when cold 0.1 M tris-HCl buffer (pH 7.5) is next added (0.25 vol for every volume of initial cell suspension). All of the following are cold-room procedures unless otherwise indicated. The precipitate is stirred in the buffer for 15 min, using a magnetic stirrer, and then allowed to stand for 5 min. Even though the tRNA component of the precipitate is completely soluble under these conditions, the overall result is often not a true solution but an opalescent-appearing, colloidal-like suspension (especially noticeable in large preparations) containing no settled particles. If particles are present, they should be removed by a brief centrifugation. The suspension is then added, with stirring, to a batch of the anion exchanger DEAE-cellulose (1 g of DEAE-cellulose per 75 g of E. coli cells) that has been equilibrated with 0.01 M tris-HCl buffer; the stirring is allowed to continue for 4 hr to insure complete binding of nucleic acids to the anion exchanger. [The DEAE-cellulose used for this procedure should have a low ion capacity, somewhere within the range of 0.7-1.0 meq/g. Depending on its source and purity, it may be necessary to precycle (wash) before equilibration. Both techniques are comprehensively explained in Whatman Technical Bulletin IE2 and by Staehelin [18]. During equilibration, all "fines" should be removed from the DEAE-cellulose by allowing the suspension to settle for 15 min.] The DEAE-cellulose is allowed to settle (mild centrifugation can be used, 1000-2000 X g for 10 min) and the supernatant, which contains unbound proteins and polysaccharides,

10. Transfer RNA

is removed and discarded. The DEAE-cellulose is washed four times (it is resuspended with stirring for approximately 10 min and allowed to settle) in 0.1 M tris-HCl (pH 7.5) (the volume used each time is the same as the volume used for suspension of the ethanol precipitate) to remove any adsorbed material. Washed DEAE-cellulose is suspended in a solution of 1 M NaCl buffered with 0.25 M tris-HCl at pH 7.5 (same volume as above) and stirred again for 4 hr to release the bound crude tRNA. Any rRNA, most rRNA fragments, and DNA fragments of a comparable size that are bound to the DEAE-cellulose are not removed under these conditions. The supernatant containing the crude tRNA is collected after mild centrifugation (1000-2000 X g for 10 min); 2.5 times its volume of cold 95% ethanol is slowly added to the supernatant, and the precipitate is allowed to form under the same storage conditions as for the first ethanol precipitation. Since this precipitate is smaller in amount and less flocculent than the first, it is collected by centrifugation at 17,000 X g for 30 min. The precipitate is washed twice with a convenient volume of 80% ethanol (it is suspended and recentrifuged) and once with 95% ethanol. It is then dried, this time completely, in the apparatus described earlier. It can be stored in the dry state (in a container with a silica gel desiccant) in a freezer at -20°C for several years without loss of amino acid acceptor activity. [As a measure of concentration, we ordinarily employ: 1 mg/ml at 260 nm (1-cm light path) = 20 A units. The ratio of A_{260} to A_{280} should be between 1.9 and 2.1. Lower ratios indicate protein contamination; higher ratios often indicate the presence of phenol.] The yield of crude tRNA is usually of the order of 0.1% of the starting weight of E. coli cells. Contamination with protein, polysaccharides, and DNA is less than 1% for each. [Excess protein can be removed by reextraction with phenol; polysaccharide can be removed by treatment with the KH_2PO_4-K_2HPO_4-2-methoxyethanol system described by Kirby [16]; to obtain two phases this system must be prepared and used at a temperature of 20°C; at normal room temperature (25°C) only one phase is formed, and at 4°C the KH_2PO_4 is not completely soluble; DNA can be removed by treatment with DNase as described in Section II,B.] Their presence can be determined quantitatively by using the Folin-Lowry procedure [19] for proteins with bovine serum albumin as the standard; the anthrone reaction [20] for polysaccharides (after acid hydrolysis of a dialyzed alkaline digest of the crude tRNA); and the diphenylamine reaction [21] for DNA.

In certain circumstances, such as when very small numbers of labeled cells are being used, it may be desirable to forego the use of DEAE-cellulose and the purity thus obtained in order to save time and, more importantly, to reduce the number of manipulations with their concomitant loss of material. This is accomplished (see also Diagram 2) by suspending the partly dried precipitate after the first addition of ethanol in a convenient volume of cold (4°C) 1 M NaCl and stirring in a cold room for 4 hr. The 1 M NaCl extract is collected by centrifugation at 17,000 X g for 30 min and the crude tRNA is precipitated by the addition of 2.5 vol of 95% ethanol. It is washed and dried as already outlined. It is often advisable when using only the 1 M NaCl extraction procedure to make a second phenol extraction of the aqueous phase (1-2 hr) before the first ethanol precipitation.

```
| Escherichia coli |
        |
        Thaw at 4°C in water
        Add 1.2 vol of water-saturated phenol
        Extract 2 hr
        Centrifuge; discard phenol layer and interface material
| Aqueous phase |
        |
        Add an equal volume of water-saturated phenol
        Reextract 2 hr
        Centrifuge; discard phenol layer and interface material
| Reextracted aqueous phase |
        |
        Precipitate as in Diagram 1
| Precipitate |
        |
        Suspend in cold 1 M NaCl (4°C)
        Let stir 4 hr
        Centrifuge and collect supernatant
| 1 M NaCl extract |
        |
        Precipitate as in Diagram 1
| Precipitate |
        |
        Wash as in Diagram 1
| Crude tRNA |
```

Diagram 2. The preparation of crude tRNA from whole cells by extraction with 1 M cold NaCl.

B. From Disrupted Cells

When it is desirable to retain various cell fractions for other purposes, crude tRNA can be prepared from the "105,000 X g" supernatant (see Diagram 3A), which is known to contain the vast

---

Escherichia coli

   Grind with 2 times the cell wet weight of alumina
   Suspend thick paste in 3 times its vol of 0.01 M
     magnesium acetate
   Add DNAse (2 mg per ml of suspension)
   Continue mixing for 1-2 min
   Centrifuge and collect supernatant
   Recentrifuge and collect supernatant
   Ultracentrifuge for 4 hr
   Remove upper one-half to two-thirds of supernatant;
     save lower third and pellets

105,000 X g supernatant

   Phenol extraction, and so on, as in Diagrams 1 or 2

Crude tRNA

A

Ribosomal fraction

   Suspend 105,000 X g pellets obtained from above in
     0.01 M tris (pH 7.4)-0.01 M magnesium acetate-0.5%
     sodium dodecyl sulfate

   Phenol extraction, and so on, as in Diagram 1 only

Crude tRNA

B

Diagram 3. The preparation of crude tRNA
from cell fractions;
(A) 105,000 X g supernatant, (B) Ribosomal fraction.

majority of the cell's tRNA; the remainder of the RNA is found bound to the ribosomes.

A given wet weight of cells (fresh or frozen) is added to a precooled mortar (preferably wedgewood) and ground with twice their weight of washed alumina (which may be added in portions) until a thick paste accompanied by a distinctive crackling sound is obtained (see also Chapter 6). The paste is suspended in 3 times its volume of buffer [0.01 M tris (pH 7.4), 0.01 M magnesium acetate], DNase is added (2 µg per milliliter of suspension), and mixing is continued for 1-2 min. (This treatment is advisable because significant amounts of DNA fragments may be released when cells are disrupted in this fashion. The DNAse used is DNAse I, electrophoretically purified to remove RNAse, and is obtained from Worthington Biochemical Corporation.) The resulting white suspension is centrifuged at 6000 X g for 15 min (to remove alumina and some cell debris); the supernatant is decanted and recentrifuged at the same speed for 30 min to sediment the remaining cell debris. The brownish-looking supernatant is decanted and centrifuged (in a preparative ultracentrifuge) at 105,000 X g for 4 hr. Only the upper one-half to two-thirds (to avoid ribosomal subunits) of the clear supernatant is carefully drawn off and used for the preparation of crude tRNA by the methods described in Section II,A. The remainder of the supernatant can be decanted and saved along with the pellets (ribosomal fraction).

The tRNA bound to the ribosomal fraction can be extracted (see Diagram 3B) in the manner described for whole cells with the following modifications. The buffer used for suspending the ribosomes [0.01 M tris (pH 7.4), 0.01 M magnesium acetate] should contain 0.5% sodium dodecyl sulfate prior to the phenol extraction step. (Only recrystallized material should be used; a highly purified preparation listed as sodium lauryl sulfate, M.A. can be obtained from Mann-Schwartz. Because of its "detergent" properties, it is used here as an adjunct to phenol, to help disrupt RNA-protein bonds and to inhibit nuclease activity. For a review of various substances that can be used in conjunction with phenol under various extraction conditions, see the review by Kirby [22].) Also, because of the preponderance of high-molecular-weight RNA, the DEAE-cellulose treatment must be included. The product thus obtained consists of almost equal amounts of 5 S RNA and tRNA, along with some "ribosomal-like" RNA.

III. FURTHER PURIFICATION

The material obtained by the isolation procedures described in Section II was referred to as <u>crude</u> tRNA. The reason for this designation is that although there is little or no contamination of the preparation with proteins, polysaccharides, or DNA, it may contain up to 25% of RNAs other than tRNAs [12]. The relative amounts and kinds of contaminating

RNAs vary depending on the stage of the growth cycle of the cells, length of time of phenol extraction, whether DEAE-cellulose or 1 M NaCl extraction is used, and whether a whole-cell preparation or a cell fraction is chosen as starting material (see Figure 1).

Sephadex gel filtration can be used to evaluate quantitatively the contaminating RNAs while separating them from the tRNA. [It is necessary to confirm the homogeneity of material obtained in a purification step by at least one other independent method. Polyacrylamide gel electrophoresis, employing a procedure such as Loening's [24], can be used for this purpose. It is an especially advantageous method in that only small amounts are needed for monitoring the purity (and confirming the identities of the various RNAs) in both the crude tRNA preparation and the Sephadex-purified fractions.] The gel-filtration effluent profiles shown in Figure 1 demonstrate that four elution regions that contain RNA are usually obtained (sometimes a non-nucleic acid-containing peak is found in the inclusion volume of the column) with essentially all of the tRNA in elution region 4. Elution region 3 consists of 5 S RNA; region 2 contains an RNA having a molecular weight of about 54,000 [23]; region 1 (which emerges in the void volume) is least characterized and probably consists of a mixture of fragments and, possibly, intact chains of ribosomal and mRNAs.

Sephadex G-100 is the pore-size used for this separation.

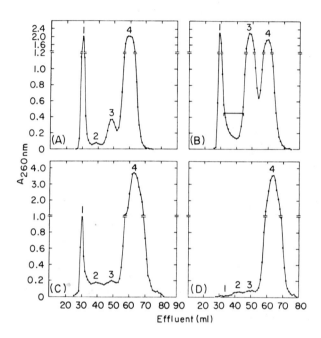

Fig. 1. Sephadex G-100 effluent profiles (the identity of the material in each elution region was confirmed by disc electrophoresis) (0.9 cm X 150 cm column) of crude tRNA prepared from (a) whole cells by 1 M NaCl extraction method; note that elution region 1 is greatly reduced if DEAE-cellulose is used; (b) ribosomal fraction by DEAE-cellulose method (elution region 2 is not found in the ribosomal fraction); (c) 105,000 X g supernatant by 1 M NaCl method; (d) 105,000 X g supernatant by DEAE-cellulose method.

10. Transfer RNA

[Extensive literature is available from the manufacturer of Sephadex (Pharmacia, Uppsala, Sweden) describing its properties, preparation for gel filtration, and so on. We discuss only the modifications employed to obtain the desired separation.] A given weight of gel powder is added slowly, with stirring (to prevent clumping), to 100 times its volume of 1 M NaCl (this relatively concentrated salt solution is used as the column eluant to minimize intermolecular associations among the various RNAs during the gel filtration) at room temperature; this column is run at room temperature. Slow stirring (to prevent shearing of gel particles and production of more fines) is continued for 1-2 hr until all of the Sephadex is well suspended. The gel is then allowed to complete its swelling by remaining in the 1 M NaCl for at least 18 hr. The settled gel is gently resuspended and allowed to settle for 5-7 min; the supernatant is drawn off along with all fine particles still in suspension. More 1 M NaCl is added, and the resuspension and decantation is repeated (usually five times) until no more fines remain in the supernatant after the 5-7 min settling time. Their removal is necessary to insure proper column flow rates. The columns used are all 150 cm in length and 0.9, 2.5, or 5 cm in diameter. If they contain sintered-glass supports (which should be of the most porous glass available so as not to interfere with flow rates), they should be so constructed that very little dead space exists between the sintered-glass support and the column outlet (to prevent excessive mixing of effluent and resultant loss of resolution).

All columns are prepared in the same way. The column is filled with column eluant (1 M NaCl), and glass beads (thoroughly washed; 3 mm for 0.9-cm column; 6 mm for 2.5- and 5-cm columns) are added until the sintered-glass support is uniformly covered by two layers; the glass beads prevent gel particles from coming into contact with the abrasive surface of the sintered glass, fragmenting, and consequently clogging the pores in the glass. The column eluant is drained so that about 15 cm remain, and the outlet is closed. A small amount of gel suspension or slurry is added; when this has settled to form a layer about 5 cm high, the outlet is opened. The 1 M NaCl is allowed to flow out; as the gel settles, more slurry is added. As the column packs, the supernatant is aspirated from the top by means of a syringe with long plastic tubing. A slowly rising horizontal gel surface indicates uniform packing. As the level of the gel reaches the top of the column, a column extension is added and packing is continued until the desired level is reached, usually 5 cm from the top of the column. A reservoir is connected, and the column is allowed to run overnight in order to stabilize the flow rate; during this time the gel will probably settle a little so that it will be necessary to add additional gel slurry to the top of the column. For the best resolution flow rates of the order of 18-22 ml/hr for the

10. Transfer RNA

0.9-cm-diameter column, 50-60 ml/hr for the 2.5-cm column, and 70-80 ml/hr for the 5-cm column should be obtained with the reservoir level never more than 50 cm above the top of the gel bed. The columns can be reused (after running through several column volumes of 1 M NaCl) until the flow rate begins to drop appreciably (i.e., 20%). Upper levels of sample size are 2 mg for 0.9-cm diameter, 50 mg for 2.5-cm diameter, and 1 g for 5-cm diameter. The minimum samples that may be applied are dictated by the type of detection device used and the problems inherent in attempting to recover relatively small amounts of separated material present in low concentrations. (A flow cell with an ultraviolet detection and recording device is very useful for indicating the degree of separation and estimating the relative amounts of the various components as they are being collected. Such a system may also be used to analyze a preparation of crude tRNA; as little as 0.1 mg can be applied to a 2.5-cm column for this purpose if the detector has the requisite sensitivity.) The following briefly describes the important statistics for each diameter of column. 0.9-cm column: sample volume should not exceed 1 ml (dry samples should be taken up in 1 M NaCl; samples in aqueous solution can be directly applied) and fractions no larger than 2 ml should be collected; incomplete separation of tRNA from 5 S RNA. 2.5-cm column: sample volume up to 4 ml; a forerun of about 60-70 ml can be collected (and discarded) before starting the fraction collector; 4-ml fractions should be collected; complete separation of tRNA from 5 S RNA. 5-cm column: sample volume up to 10 ml; forerun of approximately 600 ml; fraction size, 10 ml; complete separation of tRNA from 5 S RNA.

The purified tRNA can be isolated by dialyzing the pooled fractions in a cold room for 24 hr against several changes of deionized water (dialysis tubing that has been boiled in deionized water for 10-15 min should be used). The salt-free solution is then concentrated, using a rotary evaporator under reduced pressure (produced by a water aspirator), at a temperature of no more than 30°C. When a tRNA concentration of at least 0.1 mg/ml is reached, the evaporation may be stopped and the tRNA is precipitated as before with potassium acetate and ethanol. This final product consists of a mixture of acylated and nonacylated amino acid acceptor RNAs. If so desired, the tRNAs can be further deacylated by dissolving the preparation in a 0.1 M ammonium carbonate solution (adjusted to a pH of 10.2-10.5 with NH_4OH) and incubating at 37°C for 1 hr [25]. The deacylated tRNA is recovered by the addition of potassium acetate and ethanol.

The simplest way to determine biological activity is to test for the primary function of tRNAs, that is, their ability to accept amino acids (to become acylated or "charged"). A convenient method describing both the preparation of a crude activating enzyme mixture and assay conditions is given by

10. Transfer RNA

Goldstein et al. [26] (see also Chapter 9). To test the ability of a given aminoacyl-tRNA to transfer its amino acid into a polypeptide chain, one can employ the assay procedure presented by Weisblum et al. [27].

IV. FRACTIONATION METHODS

The methods for obtaining individual amino acid acceptor RNAs are too numerous and diverse to discuss explicitly in this chapter. Instead, a survey of many of those most often used is presented, along with a discussion of reasons for their use, and representative references that explain their use in detail are given.

A. Partition Systems

Countercurrent distribution (CCD) was the method first really successfully used to fractionate tRNA. Holley employed CCD to isolate an alanine tRNA from yeast [28] and subsequently determined the structure of this tRNA [29]. Essentially, in this method, one is dealing with the solubility properties of a multicomponent solute (tRNA) in a mixture of solvents; the solvents are selected so that they produce two immiscible phases and the maximum selectivity for separation. Selectivity means that the ratio of the concentration of different amino acid acceptor RNAs in the upper phase versus the lower phase varies depending on their relative solubilities in each phase [the ratio of the concentration of a substance in the upper versus the lower phase is known as its partition coefficient (K)]. The CCD apparatus may be considered a series of interconnected separatory funnels constructed in such a way that after the phases have settled the upper phase from each funnel can be decanted and transferred to the next funnel in line, with fresh upper phase added to the first funnel. Eventually, this sequence of transfers results in those amino acid acceptor RNAs with high partition coefficients (solubility favoring the upper phase) distributing at the far end of the apparatus (mainly in the upper phase), those with low K values remaining near the other end (mainly in the lower phase), and the remainder, having intermediate K values, distributing in between.

This method was the first to resolve a large number of amino acid acceptor RNAs (including isoaccepting species). An example of such a separation with E. coli tRNA is given by Goldstein et al. [26]. CCD can be used to fractionate relatively large amounts of material (gram quantities, under suitable conditions) and to separate large numbers of amino acid acceptor RNAs in milligram amounts. Unfortunately, with the two-phase systems currently employed and the usual experimental running time, the tRNAs become deacylated. This means that a specific amino acid acceptor RNA cannot be charged with a labeled amino acid and its profile determined against the general distribution of unlabeled tRNAs. Another

10. Transfer RNA

disadvantage is the comparatively high cost of CCD equipment.

To overcome these difficulties and other problems of technique, some workers turned to partition chromatography (in columns). The basic principle is the same as for CCD, that is, partition of a solute in a two-phase system. As with CCD, the lower phase (primarily aqueous in composition) is usually the stationary phase and the upper (more organic) is the mobile phase. Since partition chromatography is a column procedure, the stationary phase must be contained on some supporting substance which is packed into the column. The solute is dissolved in the mobile phase (usually some modification of the mobile phase is actually used so that the initial partition coefficient of the solute greatly favors the lower phase) and is placed on the column; then mobile phase is allowed to percolate through the column. Presumably, the tRNAs partition between the phases in a manner analogous to that occurring in CCD appartus or in a single separatory funnel. Muench and Berg [30] devised such a system which includes the addition of triethylamine, in a linear concentration gradient, to the mobile phase; the triethylamine selectively extracts different amino acid acceptor RNAs (by causing a variation in their partition coefficients in the tRNA mixture). A widely used method was devised by Weiss and Kelmers [31]. This technique employs a reverse-phase (which simply means that the more organic phase is the stationary phase and the more aqueous is the mobile phase--a reverse of the usual situation) system that utilizes ion exchange as a means of enhancing the differential solubility of the various amino acid acceptor RNAs. The ion-exchange effect is produced by having a quaternary amine dissolved in the organic (stationary) phase and a gradient of NaCl in the aqueous (mobile) phase. As the concentration of sodium ion is increased, different amino acid acceptor RNAs become increasingly more soluble in the aqueous phase (as their sodium salts) and decreasingly less soluble in the organic phase (as the quaternary salt). Both of these column partition methods have the advantage of large sample capacity and of being able to separate a large number of individual amino acid acceptor RNAs, including isoaccepting species [32]. Furthermore, the reverse-phase procedure can be used to profile specific amino acid acceptor RNAs charged with labeled amino acids, since deacylation does not occur during the fractionation [33]. Care must be taken when using any partition chromatographic method to be sure that the inert supporting material is truly inert under the conditions used; adsorption and subsequent desorption of the solute can lead to spurious results. Another source of artifacts are conditions (such as column flow rates) that result in incomplete equilibration of the solute between the stationary and mobile phases (conditions that can be rigidly controlled in the CCD procedure).

B. Miscellaneous Chromatographic Methods

An older column chromatographic method [34] useful in profiling labeled amino acid acceptor RNAs employs methylated albumin adsorbed to kieselguhr (a naturally occurring calcium-aluminum-magnesium silicate). An MAK column has the advantage of providing a relatively rapid chromatographic method. Unfortunately, the capacity of the column is low, and a larger column cannot be used without severely reducing the resolution of the aminoacyl-tRNAs. This capacity problem has been overcome by the use of silicic acid, in place of kieselguhr, as the inert support for the methylated albumin (MASA). An increase in resolution is also obtained with MASA [35].

Serum albumin is known to bind lipophilic or hydrophobic compounds. The esterification of its free carboxyl groups with methanol produces an anion exchanger (the albumin becomes a basic substance because of free amino and guanidino groups) that still retains its ability to interact with hydrophobic molecules. It is this combination of anion exchange with secondary binding forces (hydrophobic interactions), the latter being quite important, that produces the separation of tRNAs. Substitution of the hydroxyl groups of the anion exchanger DEAE-cellulose by aromatic acids also produces derivatives having increased secondary attractions for nucleic acids. So far, the benzoylated derivative (BD-cellulose) has been found to give the best results [36]. Its capacity is much higher than either methylated albumin or DEAE-cellulose itself. Preparative fractionations utilizing 5 g of tRNA have been carried out on relatively small columns with good resolution. Also, certain aminoacylated tRNAs can be purified rapidly and in good yield using BD-cellulose [37].

Two other column procedures are often mentioned in the literature; both have relatively low capacities but are useful in resolving individual amino acid acceptor RNAs. One such method employs DEAE-cellulose or DEAE-Sephadex and urea buffers [38]. The other is a form of adsorption chromatography, utilizing hydroxylapatite (a hydrated calcium phosphate gel) as the adsorbant [39].

It should be clear from the preceding discussion that there is a tRNA fractionation method to suit every need; often two or more are combined to provide the desired result (for example, the isolation of methionine tRNAs [40]).

This work was supported by Grant #HE-09011 from the National Heart and Lung Institute of the National Institutes of Health.

REFERENCES

[1] M. B. Hoagland, Biochim. Biophys. Acta, 16, 288 (1955).
[2] M. B. Hoagland, E. B. Keller, and P. C. Zamecnik, J. Biol. Chem., 218, 345 (1956).

[3] J. A. DeMoss, S. M. Genuth, and G. D. Novelli, Proc. Natl. Acad. Sci. U.S., 42, 325 (1956).

[4] P. Berg and G. Newton, Federation Proc., 15, 219 (1956).

[5] R. W. Holley, J. Am. Chem. Soc., 79, 658 (1957).

[6] R. W. Holley and J. Goldstein, J. Biol. Chem., 234, 1765 (1959).

[7] T. Hultin and G. Beskow, Exptl. Cell Res., 11, 664 (1956).

[8] F. H. C. Crick, as quoted by M. B. Hoagland, in The Nucleic Acids, Vol. 3 (E. Chargaff and J. N. Davidson, eds.), Academic Press, New York, 1955, p. 400.

[9] M. B. Hoagland, P. C. Zamecnik, and M. L. Stephenson, Biochim. Biophys. Acta, 24, 215 (1957).

[10] G. D. Novelli, Ann. Rev. Biochem., 36, 449 (1967).

[11] International Union of Pure and Applied Chemistry--International Union of Biochemistry Commission on Biochemical Nomenclature, European J. Biochem., 15, 203 (1970).

[12] T. Schleich and J. Goldstein, J. Mol. Biol., 15, 136 (1966).

[13] R. W. Holley, J. Apgar, B. P. Doctor, J. Farrow, M. A. Marini, and S. H. Merrill, J. Biol. Chem., 236, 200 (1961).

[14] R. Monier, M. L. Stephenson, and P. C. Zamecnik, Biochim. Biophys. Acta, 43, 1 (1960).

[15] A. Gierer and G. Schramm, Nature, 177, 702 (1956).

[16] K. S. Kirby, Biochem. J., 64, 405 (1956).

[17] R. W. Holley, J. Apgar, and S. H. Merrill, J. Biol. Chem., 236, PC42 (1961).

[18] T. Staehelin, in Progress in Nucleic Acid Research and Molecular Biology, Vol. 2 (J. N. Davidson and W. E. Cohn, eds.), Academic Press, New York, 1963, p. 169.

[19] O. H. Lowry, N. J. Rosebrough, A. L. Farr, and R. J. Randall, J. Biol. Chem., 193, 265 (1951).

[20] F. A. Loewus, Anal. Chem., 24, 219 (1952).

[21] Z. Dische, in The Nucleic Acids, Vol. 1 (E. Chargaff and J. N. Davidson, eds.), Academic Press, New York, 1955, p. 287.

[22] K. S. Kirby, in Progress in Nucleic Acid Research and Molecular Biology, Vol. 3 (J. N. Davidson and W. E. Cohn, eds.), Academic Press, New York, 1964, p. 1.

[23] J. Goldstein and K. Harewood, J. Mol. Biol., 39, 383 (1969).

[24] U. E. Loening, Biochem. J., 102, 251 (1967).

[25] M. Nirenberg and P. Leder, Science, 145, 1399 (1964).

[26] J. Goldstein, T. P. Bennett, and L. C. Craig, Proc. Natl. Acad. Sci. U.S., 51, 119 (1964).

[27] B. Weisblum, S. Benzer, and R. W. Holley, Proc. Natl. Acad. Sci. U.S., 48, 1449 (1962).

[28] J. Apgar, R. W. Holley, and S. H. Merrill, J. Biol. Chem., 237, 796 (1962).

[29] R. W. Holley, J. Apgar, G. A. Everett, J. T. Madison, M.

10. Transfer RNA

Marquisee, S. H. Merrill, J. R. Penswick, and A. Zamir, Science, 147, 1462 (1965).

[30] K. H. Muench and P. Berg, Biochemistry, 5, 970 (1966).

[31] J. F. Weiss and A. D. Kelmers, Biochemistry, 6, 2507 (1967).

[32] D. H. Muench and P. A. Safille, Biochemstry, 7, 2799 (1968).

[33] R. C. Gallo and S. Pestka, J. Mol. Biol., 52, 195 (1970).

[34] N. Sueoka and T. Yamane, Proc. Natl. Acad. Sci. U.S., 48, 1454 (1962).

[35] R. Stern and U. Z. Littauer, Biochemistry, 7, 3469 (1968).

[36] T. Gillam, S. Millward, D. Blew, M. von Tigerstrom, E. Wimmer, and G. M. Tener, Biochemistry, 6, 3034 (1967).

[37] O. H. Maxwell, Biochemistry, 7, 2629 (1968).

[38] J. D. Cherayil and R. M. Bock, Biochemistry, 4, 1174 (1965).

[39] R. L. Pearson and A. D. Kelmers, J. Biol. Chem., 241, 767 (1966).

[40] P. Schofield, Biochemistry, 9, 1694 (1970).

Chapter 11

PREPARATION OF COLIPHAGE RNA

Daniel Kolakofsky

Institut für Molekularbiologie
Universität Zürich
Zürich, Switzerland

I. INTRODUCTION 267

II. COLIPHAGE GROWTH 268

III. COLIPHAGE PURIFICATION 270

IV. PREPARATION OF COLIPHAGE RNA 273

REFERENCES 276

I. INTRODUCTION

Coliphage RNA (RNA from bacteriophages that attack Escherichia coli) can be used as a template for in vitro protein synthesis [1], as a template for the virus-induced replicase [2], as a source of infectious nucleic acid for spheroplasts [3], as a sedimentation marker for density gradient centrifugation [2], or as a source of guanosine-5'-triphosphate 3'-monophosphate (pppGp) [4]. Since RNA coliphages can be easily prepared in meaningful quantities, and the base sequence of their RNA will (hopefully) soon be known, coliphage RNA has been to date and remains the most readily available and best understood "bacterial" mRNA.

II. COLIPHAGE GROWTH

RNA coliphages, such as f2, MS2, R17, and Qβ, are grown in male strains of E. coli, for example, Hfr 3000 [5] and Q13 (a low in ribonuclease activity strain [6]), in the following manner. One liter of prewarmed yeast-tryptone broth [7a] in a 2-liter Erlenmeyer flask is inoculated with 5 ml of an overnight culture of E. coli, and the bacteria are grown at 37°C to a concentration of 5×10^8 cells per milliliter, with vigorous aeration in a Gyrotory (New Brunswick Scientific Company) shaker bath (approximately 3-4 hr). The cell density is most easily determined by measuring the absorbance, or more correctly the turbidity, of the culture at 650 nm. For this purpose a standard curve is prepared plotting A at 650 nm versus the cell density as measured directly in a Petroff-Hausser counter. When A is measured in a 3-ml, 1-cm-light-path cuvet, an A at 650 nm of 0.3 is generally equivalent to a cell density of 5×10^8/ml. In order to minimize contamination of the bacterial culture prior to infection, aliquots should be taken for absorbance determinations only when the turbidity of the culture is clearly visible. The bacteria are then infected, with a multiplicity of infection [moi, the ratio of

11. Preparation of Coliphage RNA

infectious phage particles or plaque-forming units (PFU) to bacterial cells] of 5-10, and the infected culture is incubated for 3 hr to allow for phage growth. At this time some of the bacteria will have lysed, as evidenced by the appearance of cell debris (e.g., long filamentous strands of DNA) in the culture flask. However, to complete cell lysis, 50 mg of egg-white lysozyme, dissolved in 5 ml of 0.1 M tris-HCl (pH 8.0) and 10 μg of DNase (the DNase need not be electrophoretically purified) are added. Incubation is continued for a further 30 min, and then 2 ml of 0.5 M EDTA and 2 ml of chloroform are added and the culture is incubated for a further 30 min. The cellular events involved in the actual lysis of the bacteria are described elsewhere [7b].

Phage grown in this fashion give titers of approximately 10^{12} PFU/ml [8]. It is, however, possible to increase the phage yield by 3 to 5 times by either of the following procedures: (1) The bacteria are infected with an moi of 0.5 instead of 5-10, and phage growth is continued for 4-5 hr; this procedure allows a subsequent round of infection to take place with phage generated in the first round. (2) When the bacterial culture has reached a cell density of 5×10^8 cells per milliliter, it is centrifuged for 5 min at 8000 X g (20°C) and the bacterial pellet is resuspended in fresh broth and incubated for 5 min at 37°C before the addition of the phage. The reasons for increased phage yield with this latter procedure are unclear.

Phage grown with either of the above modifications give titers of 10^{13} PFU/ml and a yield of approximately 30 mg of phage per liter after purification (the yield is slightly lower with Qβ).

The phage lysate is then cooled to 0°C in ice and can, if necessary, be left overnight at 4°C. At this stage an aliquot of the lysate is set aside for future phage growth; a small droplet of chloroform is always added to the lysate to prevent bacterial contamination during storage.

The above procedure can also be used for the preparation of phage containing radioactive RNA. ^{3}H- or ^{14}C-labeled phage is best prepared by adding radioactive uracil or uridine to the culture (yeast-tryptone broth) 10 min after infection. ^{32}P-labeled phage is prepared by adding radioactive inorganic phosphate as with the ^{14}C- or ^{3}H-labeled precursor, except that a low concentration of phosphate should be used in the medium for bacterial growth (e.g., tryptone medium without yeast but supplemented with thiamine [7a]) so that the specific activity of the inorganic phosphate is not too diluted.

III. COLIPHAGE PURIFICATION

The phage purification procedure is essentially that of Strauss and Sinsheimer [9].

The phage is first precipitated by adjusting the lysate to 50% saturation with solid $(NH_4)_2SO_4$ (300 gm/liter). After the lysate is held 1 hr at 0°C to insure complete precipitation of

11. Preparation of Coliphage RNA

the phage, it is centrifuged for 20 min at 10,000 X g (4°C), the supernatant is discarded, and the precipitate is dissolved in 20 ml of 50 mM tris-HCl (pH 7.5)-100 mM NaCl-5 mM EDTA (TNE). Since this precipitate contains a good deal of cellular protein and DNA, which is not completely degraded, it is difficult to dissolve; repeatedly drawing the suspension through a 10-ml pipet is helpful in breaking up lumps and solubilizing it. The solution is then divided equally into two 30-ml Corex centrifuge tubes and allowed to cool to 0°C (5 min on ice); 10 ml of ice-cold Freon ($CFCl_3$) are added to each tube, the tubes are stoppered with Saran Wrap and a Caplug (an aluminum top that fits over the tube), and then vigorously shaken for 5 min in an ice-filled plastic glove (Freon boils at 11°C). The Freon and aqueous phases are separated by centrifugation (10 min at 17,000 X g, 0°C), and the upper (aqueous) phase containing the phage is carefully removed without disturbing the disc of cellular protein that has been denatured by the Freon and is found at the interface.

The phage is then further purified from cellular protein and nucleic acid and also concentrated by pelleting in an ultracentrifuge (2 hr at 105,000 X g at 4°C). The supernatant is discarded and the glassy (somewhat brownish) phage pellet is dissolved in 3 ml of TNE with the aid of a pipet and centrifuged for 10 min at 17,000 X g (4°C) to remove insoluble debris. The supernatant is then transferred to a weighed, cellulose nitrate, ultracentrifuge tube (1/2 X 2 in.) and additional TNE is added to give a total solution weight of 4.0 g; 2.40 g of CsCl are then added to the tube and dissolved in the phage solution with the aid of a Pasteur pipet. The density of the solution should now be 1.40 gm/cm^3, and this can be checked by weighing a 100-μl aliquot in a Pedersen-type micropipet whose exact volume has been previously determined by weighing the same pipet filled with water. Balance counterweight tubes (±0.1 g) are prepared with $ZnCl_2$ solutions of equal density (528 mg/ml [10]), and all centrifuge tubes are layered with 0.2 ml of paraffin oil to prevent evaporation during the prolonged equilibrium density centrifugation. The phage is then banded at its bouyant density (1.44-1.47 gm/cm^3 [11]) by centrifugation for 16 hr at 33,000 rpm (15°C) in a SW 39 Spinco rotor. (If small swinging-bucket rotors are not available, CsCl banding of the phage can also be performed in fixed-angle rotors.) Upon centrifugation the virus forms a clear refractive band (except in the case of Qβ, which forms a white, partially precipitated band) which is clearly visible in the lower half of the centrifuge tube. Occasionally, a second less dense and more diffuse phage band is also obtained, but this upper band is not included in the phage preparation since it contains partially degraded RNA. The upper portion of the CsCl gradient is removed to within 1 mm of the phage band with a Pasteur pipet and discarded; then the phage is similarly removed, care being taken to keep the tip of the pipet at the very top of the phage band. The phage should now be completely separated from contaminating proteins, which appear as a glassy film at the top of the gradient, and cellular nucleic acid, which is found as a pellet at the bottom of the tube.

The pure phage solution (0.5-0.7 ml) is then desalted (of CsCl) by gel filtration on a 15-ml column of Sephadex G-50 in 0.15 M NaCl-0.015 M sodium citrate (pH 6.5) (this solution is referred to as standard saline citrate, SSC); the excluded phage fractions are located by their bluish opalescence. Alternatively, the CsCl can be removed by extensive dialysis against the same buffer (SSC). The concentration of the phage solution can then be determined by it absorbance; an A_{260} nm of 8.03 is equivalent to a phage concentration of 1.0 mg/ml [9]. The phage is stored as a concentrated solution (>1 mg/ml) at -20°C; it is stable indefinitely under these conditions if it is not repeatedly frozen and thawed.

IV. PREPARATION OF COLIPHAGE RNA

RNA coliphages consist solely of single-stranded viral RNA surrounded by a protein coat; thus their RNA is isolated by simply deproteinizing the phage. This deproteinization can be accomplished by several means, for example, treatment with acetic acid [12] or guanidine-HCl [13]. However, phenol extraction has been the method of choice for many years [14,15], and the following procedure deals only with this method (see also Chapter 10).

Phage that has been well purified from contaminating nucleases and cellular RNA should be used. Procedures that utilize banding in CsCl as the final purification step best accomplish this end. To insure intactness of the RNA during its isolation, all glassware to be used should be (1) freed of heavy-metal cations (which catalyze the hydrolysis of RNA) by treatment with dilute (0.5%) EDTA solutions, and (2) made nuclease-free by either heating overnight in a 140°C oven, or by soaking for 10 min in a 15% solution of H_2O_2 and then rinsing well with distilled water, or by treatment with diethylpyrocarbonate [16]. In addition, all solutions to be used should be autoclaved to destroy nucleases.

For phenol extraction a solution of phage that contains 1-10 mg/ml in SSC and that is adjusted to contain 0.2% sodium dodecyl sulfate (SDS) is added to a centrifuge tube. (The detergent SDS is used in the deproteinization as it inhibits nucleases and, being anionic, does not complex with the RNA. The potassium salt of this detergent is poorly soluble, and hence buffers containing potassium ions should be avoided.) As an extra safeguard against nucleases, the addition of 1/10 of the virus weight of Macaloid [a clay that tends to adsorb and inactivate RNase (National Lead Company, Houston, Texas); added as a 50 mg/ml suspension] that has been autoclaved in distilled water is helpful [17]. An equal volume of redistilled phenol equilibrated with SSC is then added, and the tube is vigorously

11. Preparation of Coliphage RNA

shaken for 3 min at room temperature. (A Saran Wrap-covered rubber stopper or Caplug is a good way to close the tube during this operation.) The phases are then separated by centrifugation (5 min at 1000 X g) at room temperature and the upper (aqueous) phase is transferred to a second centrifuge tube for another extraction with phenol as before; the extraction is repeated two more times. A special note of caution should be mentioned at this stage. The phenol-water interface is likely to contain nucleases and under no circumstances should any phenol be removed with the aqueous phase. If more quantitative yields are required, most of the RNA-containing aqueous phase can be recovered by layering additional SSC on, thereby "rinsing," the phenol phase.

One-tenth volume of 2 M sodium acetate (pH 5.2) is added to the aqueous phase to lower the solubility of the RNA, which is then precipitated by the addition of 2 vol of cold ethanol. After 2 hr at $-20°C$ to insure complete precipitation, the RNA is recovered by centrifuging for 10 min at 20,000 X g ($4°C$). The RNA pellet is dissolved in distilled water; the remaining phenol is removed by repeated extraction with peroxide-free ether (only ether from unopened containers or those stored at $-20°C$ to prevent peroxides from forming should be used), and the ether is then removed by blowing nitrogen over the surface of the solution. The RNA preparation is now pure except for trace amounts of SDS, which can be removed by repeating the ethanol precipitation step.

Alternatively, after the first ethanol precipitation step, the RNA is centrifuged through a 5-20% sucrose gradient containing 50 mM tris-HCl (pH 7.5) and 5 mM EDTA and is recovered from the peak fractions by ethanol precipitation as before. This alternative procedure has the advantage not only of removing all the contaminating phenol and SDS but of selecting only full-length RNA (27-30 S) in case of any breakdown of the phage RNA during its isolation.

The yield in both the above procedures is approximately 80%. The RNA can be stored either as an ethanol precipitate or in 0.1 mM EDTA at -20°C, and its concentration is determined by its absorbance; the A_{260} of a 1 mg/ml solution at neutral pH is equal to 25.1 [9].

REFERENCES

[1] Cold Spring Harbor Symp. Quant. Biol., 34, 651 ff (1969).

[2] Cold Spring Harbor Symp. Quant. Biol., 33, 1 ff (1968).

[3] J. H. Strauss, Jr. and R. L. Sinsheimer, J. Virol., 1, 711 (1967).

[4] R. Roblin, J. Mol. Biol., 31, 51 (1968).

[5] T. Loeb and N. D. Zinder, Proc. Natl. Acad. Sci. U.S., 47, 282 (1961).

[6] W. Gilbert, unpublished results; W. T. Hsieh and J. M. Buchanan, Proc. Natl. Acad. Sci. U.S., 58, 2468 (1967).

[7a] M. A. Billeter and C. Weissmann, in Procedures in Nucleic

Acid Research, Vol. 5 (G. L. Cantoni and D. R. Davies, eds.), Harper, New York, 1967, p. 498.

[7b] D. C. Birdsell and E. H. Cota-Robles, J. Bacteriol., 93, 427 (1967).

[8] For the methodology of virus determination, see M. H. Adams, Methods Med. Res., 2, 1 (1950) and Ref. 7a.

[9] J. H. Strauss, Jr. and R. L. Sinsheimer, J. Mol. Biol., 7, 43 (1963).

[10] Handbook of Physics and Chemistry, Chemical Rubber Publishing Company, Cleveland, Ohio, 1970.

[11] T. Nishihora and I. Watanabe, Virology, 39, 360 (1969).

[12] H. Frankel-Conrat, Virology, 4, 711 (1957).

[13] R. A. Cox, in Methods in Enzymology, Vol. 12B (L. Grossman and K. Moldave, eds.), Academic Press, New York, 1970, p. 120.

[14] A. Gierer and G. Schramm, Z. Naturforsch., 138 (1956).

[15] R. F. Gesteland and H. Boedkter, J. Mol. Biol., 8, 496 (1964).

[16] F. Solymosy, I. Fedorcsak, A. Gulyas, G. L. Farkas, and L. Ehrenberg, European J. Biochem., 5, 520 (1968).

[17] W. M. Stanley, Jr. and R. M. Bock, Biochemistry, 4, 1302 (1965).

Chapter 12

RIBOSOMAL RNA

Frederick Varricchio

Department of Internal Medicine
Yale University Medical School
New Haven, Connecticut

I.	INTRODUCTION	279
II.	BACTERIAL STRAINS AND CULTURE CONDITIONS	280
III.	PREPARATION OF CELL EXTRACTS	282
IV.	PREPARATION AND LYSIS OF E. coli SPHEROPLASTS. . . .	283
V.	PREPARATION OF RIBOSOMES	284
VI.	EXTRACTION OF RNA.	288
VII.	SEPARATION OF RNA.	290
VIII.	POLYACRYLAMIDE GEL SEPARATION OF RNA	291
IX.	CHARACTERIZATION OF RNA.	300
	REFERENCES	312

I. INTRODUCTION

Approximately 80% of the total RNA of Escherichia coli is rRNA. This RNA, along with about 50 proteins, forms the 30 and 50 S (S = Svedberg units, named after the developer of the ultracentrifuge; Svedberg units are now used to refer to

Copyright © 1971 by Marcel Dekker, Inc. No part of this work may be reproduced or utilized in any form or by any means, electronic or mechanical, including xerography, photocopying, microfilm, and recording, or by any information storage and retrieval system, without the written permission of the publisher.

RNAs according to their sedimentation rate in the ultracentrifuge) ribosomal subunits. The RNA of the 30 S ribosome, referred to as 16 S RNA, has a molecular weight of 550,000 and sediments at 16 S. The larger ribosome contains two RNAs (this is true for all species of bacteria except one; an apparent exception is Rhodopseudomonas spheroides in which the 50 S subunit contains RNA molecules of molecular weight 530,000 and 420,000 and the 5 S RNA [1]), a 23 S RNA and a 5S RNA with molecular weights of 1,100,000 and 40,000, respectively. Much current research is directed at elucidating the "fine structure" of a ribosome, that is, the organization of the various components in relation to each other. The function of the RNA in these nucleoprotein complexes is unknown.

In this chapter methods for the preparation and study of rRNA are described. The emphasis is on newer methods which permit extensive studies on small amounts of highly isotopically labeled RNA. References are given to original descriptions of a method and also to more detailed discussions elsewhere. In some cases a source of supply is mentioned. This is not meant to be exclusive but only to guide the reader to a satisfactory source.

Although the study of E. coli is referred to, the methods are of general application except where the contrary is obvious.

II. BACTERIAL STRAINS AND CULTURE CONDITIONS

Literally hundreds of E. coli strains are available. From

12. Ribosomal RNA

the point of view of RNA preparation, one class of strains is of special interest: those that lack one or more of the five E. coli RNases. Several strains that lack RNase I are generally available. Two such mutants of E. coli K12 were isolated by Gesteland [2]. One of these, strain $\overline{10}$ (D10 or D1$\overline{10}$), has been most widely used. Another strain that lacks RNase I which is often used is MRE 600.

A defined medium useful for growth of many strains (except Bacillus) is the M63 medium of Cohen and Rickenberg [3] which consists of: KH_2PO_4, 13.6 g; $(NH_4)_2SO_4$, 2.0 g; $MgSO_4 \cdot 7H_2O$, 0.2 g; $FeSO_4 \cdot 7H_2O$, 0.005 g. The pH is adjusted to 7.0 with KOH and the volume brought to 1 liter.

As a carbon source, glucose, which is sterilized separately, is added to a concentration of 0.3%. The generation time of K12 strains in this medium is 90 min at 30°C. Any radioactive compound can be added to this medium for labeling studies. If the cells are to be used only for preparation of nonradioactive RNA, a richer medium containing 0.3% yeast extract and 0.3% casamino acids (Difco) can be used. The RNA content of bacteria increases with the growth rate. Enriched media are often not suitable for labeling experiments, as a certain amount of uracil, thymine, phosphate, and so on, are contained in crude preparations such as yeast extract and dilute the specific activity of the isotope.

For the preparation of larger amounts of labeled RNA of high specific activity, ^{32}P is most useful. A low-phosphate medium suitable for ^{32}P labeling contains: tris, 14.5 g; $(NH_4)_2SO_4$, 2.0 g; $MgSO_4 \cdot 7H_2O$, 0.2 g; $FeSO_4 \cdot 7H_2O$, 0.005 g; 0.1 M K_2HPO, 1 ml. The pH is adjusted to 7.5 and H_2O added to 1 liter.

This medium contains enough phosphate to support growth of E. coli to a cell density of 0.45-0.50 absorbance units measured at a wavelength of 650 nm (A_{650}/ml).

To avoid sampling the culture after addition of ^{32}P, the following protocol may be followed. A small inoculum from a culture in low-phosphate medium is transferred to fresh medium. The cell density of the new culture is followed until the growth rate is linear. (Cell density can initially be followed at 450 nm for greater sensitivity.) The ^{32}P is added and the cells are grown for a period calculated to yield a cell density of 0.40-0.45 absorbance units at 650 nm.

III. PREPARATION OF CELL EXTRACTS

Cells are generally harvested by centrifugation. A 2-min centrifugation at 15,000 X g is sufficient to pellet bacteria. Small amounts of bacterial cells can be rapidly harvested by Millipore filtration. A culture can be poured over crushed ice or frozen media to stop cell growth rapidly. Inhibitors such as chloramphenicol, NaCN, and NaN_3 can be added at the time of harvest to inhibit protein synthesis and/or energy production.

Usually cells are disrupted before extraction of RNA because extraction of whole cells preferentially yields tRNA.

For large amounts of cells, a French pressure cell (American Instrument Company) is probably the most efficient. Sonication and alumina grinding can be used with any amount of cells. It is important to consider the Mg^{2+} concentration during the process if polysomes or 70 S monomers are to be maintained; monomers dissociate into ribosomal subunits when the Mg^{2+} concentration falls much below 5×10^{-3} M.

A very efficient cell lysis technique based on the method of Godson [4] is described below (see also Chapters 1 and 5). This technique is convenient for working with small volumes, 1 ml or less of cell suspension, and radioactive contamination is easier to control. The latter is a most important consideration when working with ^{32}P.

IV. PREPARATION AND LYSIS OF E. coli SPHEROPLASTS

Cells are resuspended in ice-cold 20% sucrose containing 5×10^{-2} M Na_2HPO_4 (pH 8.5). A cell density of 5×10^9 cells per milliliter (A_{650} about 20) is convenient and is equivalent to about 50 ml of cells from a culture at $A_{650} = 0.40$ resuspended to 1 ml. Higher cell densities may be used. It is essential to ensure that the pH of the cell suspension remains at least at 8.0. The sucrose is needed to prevent premature hypoosmotic lysis during the next step. It may be omitted if one is willing to risk exposing the cell contents to EDTA.

The 0.1 M Na_2EDTA and 1 mg/ml lysozyme (dissolved in H_2O) are added to a final concentration of 0.001 M and 100 μg/ml, respectively. After 2-5 min at 0°C, $MgSO_4$ is added to 0.01 M, Brij 58 (Atlas Chemical Industries) to 0.5% (see Chapters 1 and 5), and sodium deoxycholate to 0.2%. The final volume is now twice the original volume. Within 1-2 min the suspension should clear noticeably and become viscous. To remove the DNA 20 μg/ml DNase (Worthington, electrophoretically purified) is added.

The procedure takes place in the cold in order to maintain the polysomes and to retard other enzyme activities. If the suspension does not clear, the usual reason is that the pH during the lysozyme treatment is too low. It is often possible to recover the cells by centrifugation and to begin again.

Extracts of many strains of E. coli, as well as Aerobacter and Salmonella, have been prepared by this method. A simpler procedure may be used with Gram-positive bacteria. In that case the lysozyme treatment may be performed at pH 7, and EDTA is not needed.

From here on, one must take precautions against the ubiquitous RNases. Two precautions are usual when working with RNA: (1) glassware is heated to 225°C and (2) the operator wears rubber gloves--disposable examination gloves are adequate. In addition, sterile solutions are sometimes used, particularly for column chromotography.

V. PREPARATION OF RIBOSOMES

At this juncture, the decision as to which separation technique to use depends on the quantity of material needed,

12. Ribosomal RNA

the equipment available, and personal preference. RNA can be prepared directly from the cell extract, or a preliminary separation of ribosomes may precede the RNA preparation.

To prepare total RNA, particularly in quantity, RNA can be prepared directly from a crude cell extract. However, even though some RNA sediments with the cytoplasmic membrane, purer RNA is more easily prepared if the crude extract is first centrifuged 10 min at 20,000 X g to remove whole cells, cell walls, and other debris. (The phenol extraction described below may extract some wall or membrane components as well as RNA.)

After removal of cell debris, total ribosomes may be prepared by centrifuging an extract 6 hr at 150,000 X g. rRNA can be prepared from such ribosomes. Alternatively, the 30 and 50 S ribosomes may first be separated. Usually, adequate quantities are prepared using a Spinco SW 25.1 swinging-bucket rotor for sucrose gradient centrifugation [5]. The SW 25.1 rotor accommodates three tubes of 28 ml each, and 0.5-1.0 mg of material (an A_{260} of 15 = 1 mg ribosomes [6]) can be well separated on each gradient. The separation becomes less efficient as the sample size is increased. Sucrose [a specially purified (RNase-free) grade is available from Mann Research Laboratories] is dissolved in a buffer containing 0.01 M $MgSO_4$, 0.005 M tris-HCl (pH 7.5), and 0.060 M NaCl. (The reader should be aware that it is not certain which monovalent cations should or should not be used [7] if one wants to preserve the in vivo ribosome distribution.)

To prepare a 15% sucrose solution, for example, 15 g of sucrose are added to 85 ml of buffer. The Buchler Company gradient-forming device is as good as any commercially available (see Ref. 5, p. 135; see also Chapters 1 and 5). Gradients of 10-20% or 15-30% sucrose are used for separating ribosomal subunits. The lone requirement is that the sample be lighter than the top of the gradient, that is, it must float. Note that the final sucrose concentration in the cell lysis method given above is 10%, but the extract can be further diluted. If a 15-30% gradient is used in a 25.1 rotor, centrifugation should be 15 hr at 25,000 rpm. Gradients are fractionated by puncturing the bottom of the tube in any of a variety of devices [available from Beckman Instruments, Instrument Specialties Company (see Chapter 5)].

The separation that can be expected is shown in Figure 1. The dotted line indicates that lighter material trails into the preceding heavier material. This must be considered when choosing fractions to pool if the purest possible material is required. In the example the purest 50 S ribosomes would be obtained by pooling tubes 9 to 11, and the purest 30 S ribosomes by pooling tubes 15 and 16. The ribosomes from such a pooling may be collected by centrifugation at 150,000 X g for 6 hr, or by precipitation with 1 vol of ethanol in the presence of 10^{-2} M

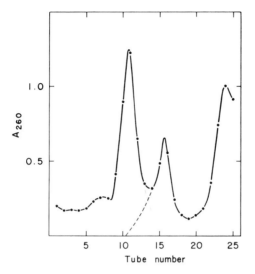

Fig. 1. Separation of ribosomal subunits by centrifugation on a sucrose gradient. One milliliter of a cell lysate was layered on a 15-30% sucrose gradient, and centrifuged 16 hr at 24,000 rpm in a SW 25.1 rotor. Fractions were collected from the bottom of the tube; according to the usual nomenclature, the first fraction (i.e., the bottom) is fraction no. 1.

$MgSO_4$ [8]. The RNA can also be extracted directly from the pooled fractions. Following the methods given, one would obtain the ribosomal subunits per se. (See Schlessinger and Apirion [9] for a discussion of the different types of ribosomes.) If the maximum amount of ribosomal subunits is to be obtained from an extract, the Mg^{2+} concentration in the gradient buffer should be reduced to 10^{-4} M Mg^{2+} or the Mg^{2+} concentration of the extract should be reduced to 10^{-3} M or less.

With zonal rotors 0.5 to 1 g quantities of a ribosomal subunit can be prepared. Conditions for separating ribosomes with the use of a zonal rotor have been published [10].

VI. EXTRACTION OF RNA

In order to extract RNA from a ribosome pellet, the pellet is first resuspended in H_2O or 10^{-2} M $MgSO_4$. The suspension to be extracted is then made 0.5% in sodium dodecyl sulfate (SDS), and an equal volume of H_2O-saturated phenol containing 0.5% hydroxyquinoline is added. (If commercial liquified phenol is not colorless, or if in doubt, it should be redistilled before use. Hydroxyquinoline is added to retard free radical formation. This is not necessarily the best procedure for all RNA extractions. For example, to extract hemoglobin messenger RNA and rRNA from rabbit reticulocytes, the newer procedure of Kirby [11] gives better results [12].) It is better if the suspension extracted is not too concentrated during this step. This mixture is agitated for 0.5-1 hr at 4°C and then centrifuged for 10 min at 1500-2000 rpm in a refrigerated centrifuge with a horizontal rotor (International Equipment Company). The phenol and aqueous layers are thus separated with denatured protein at the interface. The aqueous layer is pipetted off, about 1/2 vol of 10^{-2} M $MgSO_4$ or H_2O is added to the phenol layer, and the step is repeated. It is necessary to reextract the phenol layer to achieve a nearly quantitative yield because as much as 50% of the RNA may remain in the phenol layer after the first step.

12. Ribosomal RNA

The aqueous phase can be reextracted with phenol to insure more complete removal of denatured proteins.

To the combined aqueous layers 0.1 vol of 20% sodium acetate (pH 5.4) is added. Sodium acetate is preferred because SDS is more soluble than the K^+ salt; 2 vol of 95%, or absolute, ethanol are added and the solution is placed in the cold for at least 2 hr. It is preferable to leave it at $-20°C$ overnight. About 25 µg/ml of RNA are needed for precipitation. If less than this is available, an appropriate amount of carrier RNA is added. [Carrier RNA is best prepared from E. coli or yeast tRNA (General Biochemical). This is further purified by two cycles of phenol extraction and ethanol precipitation. Most commercial RNA is very crude and is probably not more than 50% RNA. The British Drug House yeast RNA is acceptable for some purposes but not as a carrier.] The precipitated RNA is collected by centrifugation. It can be further purified by resuspension and ethanol precipitation, one or more times, to eliminate the residual phenol usually carried over to the first ethanol precipitation. The RNA may be dried with a stream of nitrogen and further purified by washing with ether. When resuspended in buffer, the RNA should give a clear solution. Any insoluble material, usually denatured proteins, can be removed by low-speed centrifugation. The concentration of an RNA solution may be calculated on the basis of 25 A_{260} units = 1 mg RNA.

The use of several compounds has been recommended to protect RNA from degradation by the ubiquitous RNases [13]. Bentonite (a naturally occurring clay) is prepared by washing the fraction of the crude powder that sediments between 6000 X g (10 min) and 20,000 X g (15 min) in 0.1 M EDTA and then removing the EDTA by repeated washing or dialysis; it is usually added to a concentration of 5-10 mg/ml. Bentonite adsorbs RNases; it is probably best removed from the RNA solution by centrifugation. More recently, the use of 0.1% diethylpyrocarbonate (oxydiformate, Fisher Scientific Company) has been advocated for the inhibition of RNases. Higher concentrations of this compound may cause the chemical alteration of bases, but it is probably the best compound presently available for this purpose.

VII. SEPARATION OF RNA

At this point it is probably still necessary to fractionate the particular preparation of RNA to obtain its different species. Again, there are several possibilities. RNA may be purified by sucrose gradient centrifugation; 5-20% or 10-20% gradients may be used. In either case centrifugation in an SW 25-1 rotor for 15 hr at 24,500 rpm gives a separation very similar to that shown in Figure 1. Approximately 250 µg of an rRNA may be prepared in this way. If the RNA was extracted from ribosomes that were exposed to 10^{-4} M Mg^{2+} before sedimentation, thus removing the ribosome-bound tRNA, the RNA at the top of the

12. Ribosomal RNA

gradient is 5 S RNA. The RNA is then collected by ethanol precipitation of the appropriate fractions of the gradient. In such precipitations one must be wary of salt precipitation. The salt may come down with or without the RNA. Salt precipitation is favored by higher concentrations of ethanol. Less than 2 vol of ethanol are required to precipitate RNA if the RNA concentration is high enough.

The S value can be estimated on the basis of a sucrose gradient separation [14]. The S value has been found, empirically, to vary directly with the distance from the top of 5-20% or 15-30% gradients as the centrifugal force on a particle from the top to the bottom of the centrifuge tube increases as the frictional effect of the higher sucrose concentration increases. To arrive at the S value of, for example, the precursor material in a pulse-labeled ribosome preparation, 26 and 43 S ribosomes, the mature (30 and 50 S) material is taken as reference. The 17 S rRNA precursor can also be detected on sucrose gradients. It sediments about halfway down the tube, to the heavy side of 16 S RNA.

VIII. POLYACRYLAMIDE GEL SEPARATION OF RNA

The most sensitive method for separating RNA is by polyacrylamide gel electrophoresis [15]. Four basic solutions are required.

Solution A: 19 g acrylamide, 1 g N',N-methylene bisacrylamide; dissolve in H_2O to a total volume of 100 ml; filter before using; if the gel is to be scanned at 260 nm to locate the RNA bands, the acrylamides must be recrystallized from chloroform and acetone to remove ultraviolet-absorbing impurities [16].

Solution B: 6.4% DMAPN in H_2O, 3-dimethylaminopropionitrile (Eastman Organic Chemicals).

Solution C: 1.6% ammonium persulfate in H_2O.

Solution D: Buffer; 108 g tris, 9.3 g Na_2EDTA, 55 g boric acid, to give 11 liters of 10-fold concentrated buffer.

Solution C can be kept 1 month. The others are stable for several months in the cold.

To separate 16 and 23 S RNAs, acrylamide concentrations must be used at which the polyacrylamide gel formed is almost liquid. Peacock and Dingman [17] devised a double-gel system in which agarose is used to provide mechanical strength. A good system for separating RNAs in the size range 16-23 S is 2% (w/v) acrylamide-0.5% (w/v) agarose.

To prepare 40 ml of solution--enough for 12 0.5 X 10-cm gels--stir and reflux 0.2 g agarose (Seakem from Bausch and Lomb) in 27 ml of H_2O at 100°C for 15 min. Cool to 40°C and add 4 ml of 10X buffer, 2.5 ml of solution B, and 4 ml of solution A. When the temperature of the solution is 35°C, add 1.25 ml of solution C and pour the mixture into gel tubes.

It is very important that the gels have a flat upper surface. Ordinarily, this is accomplished by carefully layering 1 or 2 drops of H_2O on top of the gel solution. In this case

12. Ribosomal RNA

the gel solution is almost too dilute to support H_2O. A better method is to wait until the solution has gelled (about 1/2 hr) and then to force the gel gently out of the tube with a syringe or a rubber bulb.

A 0.5 to 1-cm piece of the gel is sliced off and the gel is sucked back into the tube. The gel is retained in the tube by Scotch Tape or Parafilm and a rubber band. Gels to be run together should be the same length. The chambers of the gel apparatus are filled with a cold 10-fold dilution of buffer and the gels are run in the cold at 200 v for 90 min. The upper chamber is connected to the negative pole.

The RNA sample is dissolved in a small volume of diluted buffer to which is added a few crystals of sucrose and bromphenol blue. A few micrograms of RNA result in a stainable band. The solution should be at least 5% sucrose so that the sample will sink to the top of the gel and not mix with the chamber buffer. Bromphenol blue is the fastest moving material on the gel and serves as a visual indication of the progress of the run.

Pyrinin Y can also be added to stain the RNA reversibly during the run. The sample, containing 100 μg of RNA, is put on a gel in as small a volume as possible, 0.1 ml or less. The sample may be put on the gel and buffer layered over it, or the sample may be put under the buffer. When the dye is almost at the bottom of the gel, the run is stopped, the gels are removed from the tubes, and they are stained. The staining solution contains: 2.3 ml glacial acetic acid, 13.5 g sodium acetate, and 0.5 g New methylene blue (Allied Chemical) per 250 ml. After 1/2 hr the dye is poured off--it can be reused--and the gel is destained by soaking in H_2O. After several months the dye may wash out of the RNA bands. It is usually possible to restain such gels.

Gels that contain a higher percentage of acrylamide are prepared by increasing the proportion of solution A and decreasing the amount of H_2O mixed with the agarose. It may be necessary to reduce the bisacrylamide concentration to obtain a gel that is not opaque at 260 nm. Above a concentration of 4% acrylamide, agarose is no longer necessary to provide mechanical strength.

To separate 5 S RNA from tRNA, 10% gels must be used. Mix 20 ml of solution A, 4 ml of solution D, 2.5 ml of solution B, 12 ml of H_2O, and 1.25 ml of solution C. A few drops of H_2O are layered on top of the gel solution. After the gel has set, the H_2O is removed from the top of the gel and 1-2 cm of a large-pore gel solution are put on top of the gel. This upper gel is used to concentrate the RNA and permits application of larger sample volumes. The upper gel is prepared from a solution containing: 0.61 g tris, 4.75 g acrylamide, and 0.25 g bisacrylamide (or ethylene diacrylate, K and K Chemicals), adjusted to pH 3.6 with HCl and made to a final volume of 100 ml with H_2O. Just before use, 0.25 ml of 0.01% riboflavin per

10 ml of stock large-pore solution is added. Polymerization occurs faster if the gel is placed in front of a fluorescent light. Again, H_2O is layered on the gel solution and removed after the gel has set.

Longer 10% gels often develop some fissures on gelling. If not too large, they usually do not interfere. Bubbles are the result of air dissolved in the gel solutions. If there is enough dissolved air to be troublesome, the gel solutions may be degassed by applying a vacuum for several minutes. Ten-percent gels are rather difficult to remove from the gel tube. It is necessary to loosen the gel from the walls of the tube with a fine needle. Forcing water between the gel and the tube with a small-gage spinal needle (25G3) is better. The gel is then forced out with a syringe.

Gels can be scanned in a Joyce-Loebel Chromoscan or a Gilford recording spectophotometer at 260 nm before staining, or at 650 nm after staining. ^{32}P-labeled RNA of high specific activity may also be detected by autoradiography, usually overnight.

A permanent record is best obtained by drying a gel slice onto filter paper. Plans for a device to slice gels lengthwise are given in Fairbanks et al. [18]. To prevent cracking 10% gels must be dried very slowly, for 12 hr or more.

The separation of 16, 17, and 23 S RNAs on a 2% gel, and of 5 S RNA on a 10% gel, is shown in Figure 2. The power of

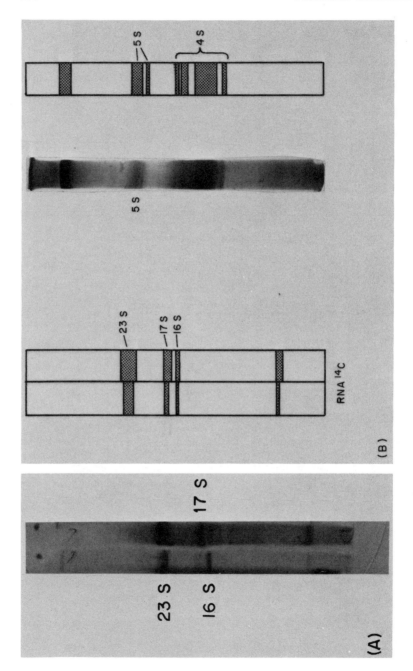

12. Ribosomal RNA

this method is well illustrated by the separation of tRNA into seven bands. Autoradiographs can be used to study rRNA precursors [19] and mRNA synthesis [20]. The precursor nature of 17 S rRNA is shown by the autoradiograph from the 2% gel of pulse-labeled RNA. The 24 S precursor RNA [19] is not clearly separated from 23 S RNA. Adensik and Levinthal identified 24 S RNA by comparing the absorbance of the gel and of its autoradiograph. The radioactivity peak is displaced slightly toward the origin. Gel slices may be counted by solubilizing slices in 30% H_2O_2 [21]. The slice is kept in a closed scintillation vial with 0.2 ml of H_2O_2 until the gel is completely liquified; it is then counted as any aqueous sample.

The molecular weight of an RNA can be calculated from its position on a gel according to Figure 3. Since migration rates

---

Fig. 2. (A) Separation of RNAs by polyacrylamide gel electrophoresis. Pulse-labeled E. coli RNA (0.08 ml), prepared as described and containing 1.0 A_{260} unit and 7500 cpm of ^{14}C, was separated on a 2.5% agarose-acrylamide gel. On the right is an autoradiograph of a gel slice exposed 28 days on x-ray film. (B) Separation of 5 S RNA on a 10% acrylamide gel. There is actually a major and a minor 5 S band. This is observed occasionally [22] and is assumed to be a result of conformation differences. Note the resolution of the tRNAs.

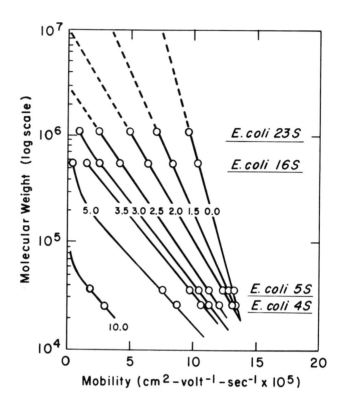

Fig. 3. Relation between the molecular weight of RNA species and their mobility in 0.5% agarose gels containing various acrylamide concentrations. The acrylamide concentrations are shown on each line. Mobility is:

$$\frac{\text{Centimeters band migrated} \times 10^5}{\text{Time of run in seconds} \times \frac{\text{Voltage across the gel}}{\text{Length of gel (centimeters)}}}$$

Reproduced from Peacock and Dingman [17] by permission of the American Chemical Society.

12. Ribosomal RNA

are rather variable, an unknown RNA should be compared to an internal standard. Figure 3 also shows the approximate maximum size of an RNA that will enter a gel of a given percentage acrylamide. Peacock and Dingman suggest that the best separation of two RNAs that are similar in size occurs on a gel that allows penetration of an RNA twice the molecular weight of the RNAs to be separated [17].

For preparative purposes gels 1 cm in diameter have been used. Slab gels (E-C Apparatus Company) may also be used on a preparative scale.

To extract RNA from a gel, the RNA band of interest is located by staining, autoradiography, or scanning at 260 nm; it is sliced from the gel and placed in a homogenizer tube with a small amount of urea [22]. The gel is homogenized (this can be done with a motor but the tube must be kept as cold as possible) and 1 ml of 0.1 M tris-0.01 M EDTA-1 M NaCl (pH 7.6) is added. The mixture is transferred to a centrifuge tube and centrifuged 5 min at 1500 rpm. The supernatant is removed and the above steps are repeated several more times. Progress of the extraction can be easily monitored by the blue stain if the gel has been stained, or with a radioactivity scanner (Victoreen Instrument Company) if the sample has been labeled with ^{32}P. The combined supernatants are extracted 1 hr with 1 vol of phenol. The phenol layer is reextracted with 1/2 vol of buffer, and 25 µg/ml of carrier RNA are added to the aqueous layers. The RNA is then precipitated with ethanol (see above). The yield is of the order of 75%.

RNA can be crudely separated by making an RNA solution (about 4 mg/ml) 1 M in NaCl in the cold. This precipitates about one-half of the rRNA.

MAK (methylated albumin kieselguhr) columns (see Chapter 10) can be used to prepare about 1 mg of rRNA. Three milligrams of RNA are put on a 2 X 25 cm column and eluted by a linear gradient, 0.2-1.1 M NaCl in 0.02 M tris-Cl (pH 6.7). The 5 S RNA is seen as a shoulder on the tRNA peak. To separate 5 S RNA from total RNA, a preliminary separation is performed on Sephadex G-100 to remove the high-molecular-weight RNA. It has also been reported that precursor RNA can be isolated on MAK columns at 27°C, using a gradient of 0.65-1.06 M NaCl [23].

5 S RNA can also be prepared by rechromatographing, on Sephadex G-100, RNA enriched in 5 S RNA. Sephadex has an advantage over MAK in that large columns can be prepared. For example, 43 mg of 1 M NaCl-soluble RNA was separated on a 3 X 182 cm G-100 column [24].

IX. CHARACTERIZATION OF RNA

A simple way to make a gross characterization of an RNA is by S value. The determination of S values requires very careful studies with an analytical ultracentrifuge. A simpler way to obtain an approximate value is by sucrose gradient centrifugation as described in Section VIII. Several formulas have been proposed that relate S value to molecular weight. One formula, which was derived in part with bacterial ribosomes by Spirin [25], is: molecular weight = $1550\ S^{2.1}$. When working with sucrose gradients, it is often useful to recall that molecular weight varies with the square of S. Since the sedimentation of an RNA in a sucrose gradient can vary with pH, Mg^{2+} concentration, and salt concentration, the estimation of an S value should be based on an internal standard. An S value is very much dependent on the shape of a molecule; thus some exceptions to the formula given above are known.

More sophisticated physical studies require experience in physical biochemistry; the interested reader should consult treatises in that field.

RNAs are also characterized by their base composition; adenine (A), cytosine (C), guanine (G), and uracil (U), and by the ratio of bases that can participate in Watson-Crick-type base pairing,

$$\frac{A + U}{G + C},$$

sometimes referred to as percent GC content.

To determine its base composition, a uniformly ^{32}P-labeled RNA sample is hydrolyzed in 0.3-0.5 M NaOH for 18 hr at 37°C.

An NaOH solution is made just before use by dissolving one pellet of NaOH in sufficient H_2O to give the desired concentration. A dried sample may be dissolved in 20 μl of 0.3 M NaOH, for example. Enough NaOH must be added so that the hydrolyzed mucleotides will not effectively lower the pH during the hydrolysis. However, the volume is kept as small as possible. Obviously, this is facilitated by having very high specific activity RNA. This can be achieved in part by minimizing volumes throughout the purification procedure so that only the minimal amount of carrier is needed. In most cases the concentration of carrier RNA is the effective RNA concentration. After hydrolysis salt can be removed by neutralizing the solution with perchloric acid, centrifuging, and removing the supernatant, which may be concentrated by evaporation with a stream of nitrogen.

The composition of major bases in each rRNA is given in Table 1. For 5 S RNA the analysis is simple since 5 S is composed of only the four major bases. These can be separated chromatographically by any <u>one</u> of the solvent systems to be mentioned. However, electrophoresis at pH 3.5 (5% acetic acid and 0.5% pyridine) gives the best separation. The 16 and 23 S RNAs contain about 1% minor or modified bases (Table 2). A two-dimensional procedure is necessary to separate most of the minor bases. The procedure that has been used is electrophoresis at pH 3.5, 1.5-2 hr at 2500 V, on Whatman no. 1 paper (40 X 60 cm)

12. Ribosomal RNA

Table 1

Base Composition of E. coli Ribosomal RNA[a]

	5 S [26]	16 S [27]	23 S [27]
Adenine	19.3	24.2	25.5
Uracil	15.9	21.3	21.0
Guanine	34.5	32.1	32.5
Cytosine	30.2	22.3	21.0

[a] Expressed as mole percent.

for the first dimension. Electrophoresis is continued until an orange G marker is almost at the end of the paper. After drying, the paper is turned 90° for descending chromotography, about 20 hr, in an isopropanol-NH_3-H_2O (70:1:29) [29] or isopropanol-HCl-H_2O (68:17, 6:14.4) (v/v/v) system [30]. Some R_U (the mobility relative to that of UMP) values in both systems for bases reported to occur in large rRNAs are given in Table 2. Success with either chromatography system is variable. Often the nucleotides move very slowly in the NH_3 system. This may be related to the temperature. In the HCl system depurination of RNA may occur because of the long exposure to 2 N HCl. The ribose phosphates produced move faster than A and G and obscure the areas to which methylated bases migrate. The butyric acid-NH_4OH system (100 parts butyric acid, 60 parts 0.5 M NH_4OH) may be best [31]. Unfortunately, R_U values of the minor bases in

Table 2

The Electrophoretic and Chromatographic properties of the Modified Components Encountered in E. coli rRNA[a]

Modified component	Symbol	Electrophoresis R_U at pH 3.5[b]	Chromatography, R_U		Source	
			Isopropanol–	Isopropanol–	16 S	23 S
Mononucleotides						
Ribothymidylic acid	Tp	0.95	1.06	—	x	x
Pseudouridylic acid	ψp	1.00	0.85	—		x
Methyluridylic acid	mUp	1.00	1.10	—		x
Guanylic acid	Gp	0.75	—	—		
N^1-Methylguanylic acid	m^1Gp	0.74	0.70	—		x
N^2-Methylguanylic acid	m^2Gp	0.68	0.73	—	x	x
N^7-Methylguanylic acid[c]	m^7Gp	0.04	0.77	—	x	x
4-Amino-5-(N-methyl)forma- midoisocytidylic acid	mGp	0.83	0.71	—	x	x
Adenylic acid	Ap	0.45	—	—		
2-Methyladenylic acid	m^2Ap	0.40	0.65	1.20		x
N^6-Methyladenylic acid	m^6Ap	0.41	0.80	1.55		x
N^6-Dimethyladenylic acid	m_2^6Ap	0.40	0.91	2.05	x	
Cytidylic acid	Cp	0.20	—	—		
5-Methylcytidylic acid	M^5Cp	0.19	0.78	—	x	x
N^4-Methyl-(2-O-methyl)cyti- dylic acid	M^4Cmp	0.26	0.82	—	x	
Dimethylcytidylic acid	m_2Cp	−0.06	0.83[d]	—	x	
	m_2^2Cp	−0.06	.			

12. Ribosomal RNA

Di- and trinucleotides					
2'-O-Methyluridylylguanylic acid	Um-Gp	1.12	—	—	x
N6-Dimethyladenylyl(N6-dimethyl)adenylic acid	m⁶ A-m⁶Ap	0.61	—	—	x
N4-Methyl-(2'-O-methyl)cytidylylcytidylic acid	m⁴Cm-Cp	0.38	0.63	1.15	x
2'-O-Methylcytidylylcytidylic acid	Cm-Cp	—	0.58	—	x
2'-O-Methylcytidylyl(2'-O-methyl)cytidylyluridylic acid	Cm-Cm-Up	0.44	—	—	"x"

aMost of the data in this table are from Fellner [28].

bR$_U$ = the mobility of the component relative to UMP.

cm⁷Gp is alkali-labile and breaks down to mGp during alkaline hydrolysis.

dR$_F$ value.

this system are not available. The butyric acid solvent may be used for some comparative purposes, even if all spots cannot be identified with certainty. Better separations are possible on thin-layer plates of cellulose if the sample is sufficiently salt-free and applied carefully enough, that is, in a very small spot. A two-dimensional separation is performed with either the butyric acid and isopropanol-NH_3 solvents or the isopropanol-NH_3 and isopropanol-HCl solvents. Some nucleotides are better separated in one system than the other. Depurination may not be a problem with solvents that contain HCl when thin-layer chromatography (TLC) is performed since a run may be completed in 2-3 hr on 20 X 20 TLC plates.

Large quantities of nucleotides, 2-5 µg (after the chromatograph is steamed in an autoclave to remove any pyridine present) may be located by shining a short-wavelength ultraviolet light (Mineral Light UVS.12, Ultraviolet Products, Inc.) on the paper or plate (microchemical amounts of material have been isolated on the thick paper, Whatman no. 3.) Trace amounts of nucleotides are located by autoradiography. If possible, at least 10 mrads are applied to the paper. The radioactivity applied to a paper is located by scanning the spot with a Geiger counter such as the Victoreen Thyac III. The paper is marked with ink, to which ^{35}S (a weak β emitter) has been added, in order to orient the autoradiograph on the chromatograph. An overnight exposure to x-ray film is sufficient to locate the

12. Ribosomal RNA

major bases. With 10 mR of starting material, a 7- to 10-day exposure is needed to locate the minor bases. Since the major bases contain 99% of the radioactivity, radiation from those compounds may obscure a minor-base spot. This may be avoided by cutting out the major-base spots after a short exposure and then reexposing the film. For quantitation the spots are cut out and counted, usually by liquid scintillation (see Chapter 6).

Column methods have been frequently used. A separation procedure using various columns to separate tRNA components has been published by Hall [32].

For several reasons, it may be of interest to determine the bases at the 5′ and/or 3′ ends of a polynucleotide. The 5′-nucleotide is unique in having two phosphate groups, pXp, after alkaline hydrolysis, while the 3′ base is released as a nucleoside.

During electrophoresis at pH 3.5, pXp migrates faster than the corresponding nucleotide. For example, in an alkaline hydrolyzate of 23 S RNA, a spot migrates in the first dimension with an R_U of 1.2. It has been identified as pGp, and is presumably the 5′ end [33]. pUp would migrate further; pCp and pAp, unfortunately migrate shorter distances and might be difficult to identify. An alternative procedure is electrophoresis at pH 5.7. At this pH, migration occurs primarily according to the number of phosphate groups. In particular, this system is useful for isolating the nucleoside, which should be quite well separated. Excellent articles on electrophoresis of nucleotides, determination of the optimal pH at which to achieve certain separations, and so on, have been written by Smith [34,35].

A procedure for labeling the 5´ end of a polynucleotide has been published [36,37]. This involves removal of the 5´-phosphate and then restoring it with a ^{32}P-labeled phosphate from γ-labeled ATP with the enzyme, polynucleotide kinase. Some difficulty may be met in purifying the kinase sufficiently from RNase activity. Also, a phenol extraction on a microscale must be made after the dephosphorylation step to remove the phosphatase [38]. A considerable investment of time is probably necessary to make this method work properly.

The identification of the 3´ end is more tractable. The 3´-ribosome can be labeled by oxidation with $NaIO_4$ followed by reduction with [^{3}H]sodium borohydride [39]. Hydrolysis then releases a labeled 3´-nucleoside or oligonucleotide.

A promising technique has been devised by Lee et al. [40] for isolation of the 3´-oligonucleotide, minus the terminal base, of a labeled polynucleotide. Such an oligonucleotide can be released by enzymic digestion with RNase T1.

To prepare an RNase T1 digest, a dried RNA sample is dissolved in 0.01 M tris-0.001 M EDTA (pH 7.4) containing 2500 units/ml of ribonuclease T1 (Calbiochem) to give an enzyme/substrate ratio of 1:20 (w/w) (10 μl of enzyme solution per μg of

RNA). After digestion at 37°C for 45 min, the reaction is stopped by placing the tube in ice.

A 20- to 50-µl aliquot of 200,000 or more Cerenkov counts per minute is added to 0.1-0.2 ml of carrier RNA (25 mg/ml). [The β radiation of ^{32}P is sufficiently strong to cause ionization of the air or H_2O through which it passes. This permits counting of ^{32}P in a scintillation counter using the ^3H setting, without addition of a scintillation mixture. Efficiency of Cerenkov counting of ^{32}P is about 25%. By this method, ^{32}P can be rapidly assayed without loss of the sample.] The carrier RNA solution should have been previously adjusted to pH 5 and the final pH of the reaction mixture should be 5. $NaIO_4$ (meta) is added to a concentration of 1.5 mg/ml and the oxidation is allowed to proceed for 30 min in the dark at 0°C. Then morpholine is added to 0.1 M and NaCl to 3 M. The pH is adjusted to 8.2 (about 1 drop of 3 N HCl per 0.1 ml). Next, 150 mg of damp aminoethyl cellulose per milliliter ml of solution [1.0 meq/g, pretreated (washed) as usual and equilibrated with 3 M NaCl-0.1 M morpholine-HCl (Aldrich Chemical Company) (pH 8.2)] are added.

This slurry is kept overnight at 0°C. During this step the oxidized 2´3´-diol, a dialdehyde, binds to the resin. The unbound RNA is removed by washing the resin with 0.1 M morpholine-3 M NaCl (pH 8.2). A Millipore filter apparatus or a sintered-glass filter can be used. When the A_{260} or counts per minute in the column washings decays to a constant value, the resin is washed three times each with 12 ml of 0.4 M $NaHCO_3$ (pH 8.2) and three times each with 12 ml of H_2O. The resin is placed in a tightly closed vessel (to minimize loss of CO_2) with 3 ml of 1 M propylamine bicarbonate (pH 7.5) per 1 ml of original reaction mixture and reacted at 37°C for 90 min. [Propylamine bicarbonate is prepared by bubbling CO_2 (dry ice) through a solution of propylamine and ether. The precipitated bicarbonate is collected by filtration, washed with ether, dried, and stored in the cold. The 1 M solution is prepared just before use and additional CO_2 is bubbled through the solution to lower the pH to 7.5. During the elimination step, one must be careful that the pH does not rise because of loss of CO_2.] This step liberates, by a β-elimination reaction, the 3´-oligonucleotide <u>minus</u> the terminal base. If required, the 3´terminal base must be identified by another method.

The resin is separated by a column technique. A Pasteur pipet plugged with glass wool works nicely. The resin is washed twice with 3 ml of triethylamine bicarbonate (pH 7.5)(Triethylamine bicarbonate is prepared by bubbling CO_2 into 30% triethylamine in H_2O for about 4 hr until the solution is pH 9) and then with 1-2 ml of H_2O. The combined column effluents containing the 3´-terminal oligonucleotide are evaporated to dryness under reduced pressure. Several evaporations may be necessary to remove all the amines. The proportion of RNA

12. Ribosomal RNA

bound to the resin and the recovery may be determined by the use of scintillation vials instead of test tubes and Cerenkov counting. The proportion of radioactivity bound to the resin gives an estimate of the number of nucleotides in the fragment.

The next step is electrophoresis at pH 1.7 on DEAE-cellulose paper, which gives a separation according to the size of the oligonucleotide. This gives an estimate of chain length, as well as purity and possible degradation during the procedure.

The fragment can be further analyzed by base analysis and the RNA sequencing methods devised by Sanger and others. These methods have been well described [26,41] and are not repeated here, except to note that the homochromatography [26] procedure has now been revised to use DEAE-cellulose TLC plates run at 60°C. The complete sequence of 5 S RNA has been reported [26], and the sequencing of the larger RNAs is well underway [42].

Sanger's fingerprinting methods may not be too useful except for identifying an unknown RNA. A T1 digest of 16 S RNA gives a map with over 100 spots. Obviously, small differences would be impossible to detect except in a fortuitous case. The difference between precursor 17 S RNA and 16 S RNA is not known. On the basis of S value, the precursor RNA may be 10% larger (about 160 bases) yet the maps of RNase T1 digests are indistinguishable [43]. However, larger 5S molecules, possibly precursors, have been identified by fingerprinting [44].

A simplified approach was used by Fellner [28] to study the methylated bases in 16 and 23S RNAs. The RNA was labeled with [^{14}C]methionine (a methionine-requiring mutant should be used to obtain efficient uptake of label) and the usual fingerprinting procedure was employed. Maps prepared from such RNA had about 10 spots. This approach may give valuable information when ribosomal RNAs are compared and in studies on their synthesis.

ACKNOWLEDGMENT

In writing this chapter, I have benefited from many conversations with my colleagues, in particular Drs. B. Forget and S. Weissman. Financial support from American Cancer Society grant E-456B and NIH grant no. 1 R01CA-10137 is acknowledged.

REFERENCES

[1] B. Marrs and S. Kaplan, J. Mol. Biol., 49, 297 (1970).

[2] R. Gesteland, J. Mol. Biol., 16, 67 (1966).

[3] G. N. Cohen and H. V. Rickenberg, Ann. Inst. Pasteur, 91, 693 (1956).

[4] G. N. Godson, in Methods in Enzymology, Vol. 12A (L. Grossman and K. Moldave, eds.), Academic Press, New York, 1967, p. 503.

[5] H. Holl, in Techniques in Protein Biosynthesis (J. Sargent and P. Campbell, eds.), Vol. 2, Academic Press, New York, 1969, Chapter 3.

[6] W. E. Hill, G. P. Rossetti, and K. E. Van Holde, J. Mol. Biol., 44, 263 (1969).

[7] L. A. Phillips, B. Hotham-Iglewski, and R. M. Franklin, J. Mol. Biol., 40, 279 (1969).

[8] A. K. Falvey and T. Staehlein, J. Mol. Biol., 53, 1 (1970).

[9] D. Schlessinger and D. Apirion, Ann. Rev. Microbiol., 23, 387 (1969).

[10] E. F. Eikenberry, T. A. Bickle, R. R. Traut, and C. A. Price, European J. Biochem., 12, 113 (1970).

[11] K. S. Kirby, in Methods in Enzymology, Vol. 12B (L. Grossman and K. Moldave, eds.), Academic Press, New York, 1968, p. 87.

[12] B. Forget, unpublished work, 1970.

[13] C. G. Rosen and I. Fedorcsak, Biochim. Biophys. Acta, 130, 401 (1966).

[14] R. G. Martin and B. N. Ames, J. Biol. Chem., 236, 1372 (1961).

[15] E. G. Richards, J. A. Coll, and W. B. Gratzer, Anal. Biochem., 12, 452, (1965).

[16] U. Loening, Biochem. J., 102, 251 (1967).

[17] A. C. Peacock and C. W. Dingman, Biochemistry, 7, 668 (1968).

[18] G. Fairbanks, Jr., C. Levinthal, and R. H. Reeder, Biochem. Biophys. Res. Commun., 20, 393 (1965).

[19] M. Adensik and C. Levinthal, J. Mol. Biol., 46, 281 (1969).

[20] M. Adensik and C. Levinthal, J. Mol. Biol., 48, 187 (1970).

[21] P. V. Tishler and C. J. Epstein, Anal. Biochem., 22, 89

[22] J. Kindley, J. Mol. Biol., 30, 125 (1967).

[23] M. R. Lamborg, Biochim. Biophys. Acta, 209, 405 (1970).

[24] M. Reynier, Ph.D. Thesis, University of Marseille, 1970.

[25] A. S. Spirin, Biokhimiya, 26, 511 (1961).

[26] G. G. Brownlee, F. Sanger, and B. G. Barrell, J. Mol. Biol., 34, 379 (1968).

[27] W. M. Stanley, Jr., and R. M. Bock, Biochemistry, 4, 1302 (1965).

[28] P. Fellner, European J. Biochem., 11, 12 (1969).

[29] R. Markham and J. D. Smith, Biochem J., 52, 552 (1952).

[30] G. R. Wyatt, Biochem. J., 48, 584 (1951).

[31] B. Magasanik, E. Vischer, R. Doniger, D. Elson, and E. Chargaff, J. Biol. Chem., 186, 37 (1950).

[32] R. Hall, in Methods in Enzymology, Vol. 12A, (L. Grossman and K. Moldave, eds.), Academic Press, New York, 1967, p. 305.

[33] B. Forget and F. Varricchio, J. Mol. Biol., 48, 509 (1970).

[34] J. D. Smith, in Methods in Enzymology, Vol. 12A, (L. Grossman and K. Moldave, eds.), Academic Press, New York, 1967, p. 350.

[35] J. D. Smith, in The Nucleic Acids (E. Chargaff and J. N. Davidson, eds.), Vol. 1, Academic Press, New York, 1955, p. 267.

[36] M. Szekely and F. Sanger, J. Mol. Biol., 43, 607 (1969).

[37] U. J. Hänggi, R. E. Streeck, H. P. Voigt, and H. G. Zachau, Biochem. Biophys. Acta, 217, 278 (1970).

[38] C. A. Marotta, personal communication (1970).

[39] U. L. RajBahandry, J. Biol. Chem., 243, 556 (1968).

[40] J. C. Lee, H. L. Weith, and P. T. Gilham, Biochemistry, 9, 113 (1970), and personal communication.

[41] F. Sanger, G. G. Brownlee, and B. G. Barrell, J. Mol. Biol., 13, 373 (1965).

[42] P. Fellner, C. Ehresmann, and J. P. Ebel, Cold Spring Harbor Symp. Quant. Biology, 35, 29 (1970).

[43] B. Pace, R. L. Peterson, and N. R. Pace, Proc. Nat. Acad. Sci. USA, 65, 1097 (1970).

[44] B. Forget and B. Jordan, Science, 167, 382 (1970).

AUTHOR INDEX

Numbers in brackets are reference numbers and indicate that an author's work is referred to although his name is not cited in the text. Underlined numbers give the page on which the complete reference is listed.

A

Abelson, P.H., 12[19], <u>60</u>
Adams, M.H., 269[8], <u>277</u>
Adamson, S., 147[18], <u>148</u>
Adensik, M., 297[19], 297[20], <u>313</u>, <u>314</u>
Ahmed, A., 68[48], <u>87</u>
Alberts, B.M., 186[23], 187[23], <u>199</u>
Albertsson, P.A., 30[51], <u>62</u> 187[24], <u>199</u>
Albrecht, J., 48[81], <u>64</u>
Alderton, G., 73[54], <u>88</u>
Algranati, I.D., 48[76, 79], 49[76], 57[76], <u>64</u>, 68 [13], <u>85</u>, 169[26], <u>176</u>
Allende, J.E., 177[4], <u>178</u> [12], <u>198</u>
Ames, B.N., 214[7], 215[7], 229[7], <u>230</u>, 291[14], <u>313</u>
Anderson, P., 16[36], 17[36], 31[36], 32[36], <u>61</u>
Anderson, W.F., 154[5], <u>174</u>
Apgar, J., 238[13,17], 257 [28,29], <u>263</u>, <u>264</u>
Apirion, D., 13[22], 16[22], 32[22], <u>60</u>, 287, <u>313</u>
Arceneaux, J.L., 68[21], <u>85</u>
Arlinghaus, R., 177[3], <u>198</u>
Aronson, A., 68[37], <u>86</u>
Aronson, J.N., 68[16], <u>85</u>
Asano, K., 68[3], <u>846</u>
Atherly, A.G., 68[7], <u>84</u>

B

Bade, E.G., 48[76,79], 49[76], 57[76], <u>64</u>, 169[26], <u>176</u>
Balassa, G., 70[52], 88
Baldwin, A.N., 215[16], 225[16], 226[16], <u>231</u>
Barrell, B.G., 303[26], 308[26], 311[41], <u>314</u>, <u>315</u>
Barrett, K., 114[21], <u>120</u>
Basilio, C., 16[38], 31[55], 32 [55, 57, 59], <u>61</u>, <u>62</u>, <u>63</u>
Bayley, S.T., 90[5,6,7,8], 98 [15], <u>109</u>
Beaud, G., 68[11], <u>85</u>
Beaudet, A.L., 203[4], 205[9], <u>211</u>, <u>212</u>
Beller, R.J., 48[73,78], <u>64</u>
Bennett, T.P., 257[26], 258[26], <u>264</u>
Beuzer, S., 257[27], <u>264</u>
Berg, P., 114[21], <u>120</u>, 205, <u>212</u>, 214[1,6], 215[9,14,15, 16], 216[6], 217[31,33], 221 [9], 222[9,33], 225[6,16,33] 226[6,16,33], 227[6], 229 [6], <u>230</u>, <u>231</u>, <u>232</u>, 236[4], <u>259</u>, <u>263</u>, <u>265</u>
Bergmann, F.H., 217[33], 220[30] 225[33], 226[33], <u>232</u>
Beskow, G., 236[7], <u>263</u>
Bickle, T.A., 288[10], <u>313</u>
Billeter, M.N., 268[7a], 270[7a] <u>276</u>

Birdsell, D.C., 269[76], 277
Bishop, H., 68[1,31], 84
Bishop, J.D., 177[1], 197
Blew, D., 204[7], 212, 261[36] 265
Bleyman, M., 68[33], 86
Bloemers, H.P.J., 222[47], 233
Böck, A., 216[25], 226[25], 229[25], 231
Bock, R.M., 216[38], 265 274[17], 277, 303[27], 314
Boedtker, H., 30[52,] 45[52], 62, 273[15], 277
Bolle, A., 2[6], 15[6], 19 [6], 23[6], 29[6]
Bolton, E.T., 12[19], 60
Bosch, L., 48[81], 64
Brawerman, G., 11[18], 15[27] 59, 60, 151[2], 154[10], 156[10], 160[15], 174, 175
Bray, G.A., 171[27], 176
Brenner, S., 128[8,9], 148
Brimacombe, R., 204[6], 212
Britten, R.J., 12[19], 129, 130[10], 148, 60
Brot, N., 178[12], 198
Brocon, J.C., 169[9], 173[9] 174[9], 175
Brownlee, G.G., 303[26], 308 [26], 311[41], 314, 315
Bruton, G.J., 216[22], 225[22] 226[22], 231
Bryan, R.N., 113[14,16], 117[14,16], 119, 120
Bubela, B., 68[10], 85
Buchanan, J.M., 268[6], 276
Burtsztyn, H., 80[56], 88

C

Calendar, R., 214[6], 216[6], 226[6], 227[6], 229[66], 230
Cammack, K.A., 3[8], 59
Campbell, L.L., 68[44], 87

Campbell, P., 285[5], 312
Cantoni, G.L., 215[9], 221 [9], 222[9], 228[51], 230, 233, 268[7a], 270 [7a], 277
Capecchi, M.R., 2[4], 6[4], 12 [4], 13[21], 14[21], 15[2], 19[34], 21[4], 22[4], 28 [48], 29[21], 41[4], 43[4], 59, 60, 61, 62, 127[7], 148, 202, 211[14], 212
Caryk, T., 202[2], 211
Caskey, C.T., 202, 205[9], 211 [11], 212
Caskey, T., 203[3,4], 211[12, 13], 212
Chae, Y.B., 156[12], 158[12], 162[18], 167[12], 168[18], 174[18], 175
Chamberlin, M., 113[19], 114 [21], 120
Chambers, D.A., 113[7], 118
Chambon, P., 68[8,39],84, 87
Chang, F.N., 68[46], 87
Changchien, L.M., 68[16], 85
Chapeville, F., 215[18], 218 [18], 231
Chargaff, E., 236[8], 246[21], 263, 264, 303[31], 308[35], 314
Cherayil, J.D., 261[38], 265
Christian, J.H.B., 90[4], 109
Clark, B.F.C., 32[58], 63
Cohen, G.N., 281, 312
Cohen, S.S., 29[49], 62
Cohn, W.E. 244, 250, 264
Cole, F.X., 220[37], 232
Cole, L.J., 228[50], 233
Coleman, G., 68[17], 85
Coll, J.A., 291[15], 313
Colowick, S.P., 6[15], 9[15], 59, 75[55], 77[55], 88,106 [18], 110, 184[22], 199, 222, [48], 233
Cooper, D., 195[26], 199
Corcoran, J.N., 17[42], 61, 68 [47], 87
Cota-Robles, E.H.,269[76], 277
Cowie, D.B., 12[19], 60

Cox, E.C., 16[37], <u>61</u>
Cox, R.A., 273[13], <u>277</u>
Craig, L.C., 257[26], 258 [26], <u>264</u>
Crick, F.H.C., 236, <u>263</u>
Cundliffe, E., 68[49], <u>87</u>

D

Davern, C., 128[9], <u>148</u>
Davidson, J.N., 236[8], 244, 246[21], 250, 308[35], <u>263</u>, <u>264</u>, <u>314</u>
Davies, D.R., 215[9], 221[9], 228, [51], <u>230</u>, <u>233</u>, 268 [7a], 270[7a], <u>277</u>
Davies, J., 16[33, 36, 39], 17[36,40], 30[53], 31 [33,36,53], 32[36,60], 42[40], 44[40], <u>61</u>, <u>62</u>, <u>63</u>, 68[14], <u>85</u>
Davis, B.D., 13[23], 14[23], 15[33], 16[23,36], 17 [33, 36], 26[23], 30[53], 31[36,53], 32[36], 35[23], 42[23,63,64], 43[23], 44 [23,63], 46[23], 47[23], 48[73,74,75,77,78], 49 [84,85], 50[84], 52[88], 53[84], 54[84], 57[84], 58[74], <u>60,61,62,63,64,65</u>, 169[25], 174[25], <u>176</u>
Delorenzo, F., 214[7], 215 [7], 229[17], <u>230</u>
DeLuca, M., 221[44], <u>232</u>
DeMoss, J.A., 236[3], <u>263</u>
Deutscher, M.P., 68[8,39], <u>84</u>, <u>87</u>
DeVries, J.K., 113[3,4], <u>118</u>
Dewey, K.F., 161[16], 162 [17], 165[22], 172[17], <u>175</u>
Dieckmann, M., 217[33], 220 [33], 225[33], 226[33], <u>232</u>
Dingman, C.W., 292[17], 298, 299[17], <u>313</u>
Dische, Z., 246[21], <u>264</u>
Doctor, B.P., 238[13], <u>263</u>

Doi, R.H., 68[1,2,18,19,30,31, 50,51], <u>84</u>,<u>85</u>,<u>86</u>,<u>87</u>
Doniger, R., 303[31], <u>314</u>
Doty, P., 169[9], 173[9], 174 [9], <u>175</u>
Douthit, H.A., 68[36], <u>86</u>
Dubnau, D., 68[32], <u>86</u>
Dubnoff, J.S., 163[19], 167[19], 169[19], <u>175</u>

E

Ebel, J.P., 311[42], <u>315</u>
Ehrenberg, L., 274[16], <u>277</u>
Ehresmann, C., 311[42], <u>315</u>
Eigner, E.A., 221[42], <u>232</u>
Eikenberry, E.F., 288[10], <u>313</u>
Eisenstadt, J.M., 11[18], 15[27] <u>59</u>, <u>60</u>, 151[2], 154[10], 156 [10], <u>174</u>, <u>175</u>
Elson, D., 303[31], <u>314</u>
Engelhardt, D.L., 6[14], 16 [14,34a], 21[14], 22[14], 28[14], 41[34a], 42[34a], 45[34a], 47[34a], <u>59</u>, <u>61</u>
Epstein, C.J., 297[21], <u>314</u>
Everett, G.A., 257[29], <u>265</u>
Ertel, R., 178[12], <u>198</u>

F

Fairbanks, G., 295[18], <u>313</u>
Falvey, A.K., 147[19], <u>149</u>, 287[18], <u>313</u>
Farkas, G.L., 274[16], <u>277</u>
Farr, A.L., 98[14], <u>110</u>, 246 [19], <u>264</u>
Farrow, J., 238[13], <u>263</u>
Favelukes, G., 177[3], <u>198</u>
Fedorcsak, I., 274[16], <u>277</u>, 290[13], <u>313</u>
Feinsod, F.M., 68[36], <u>86</u>
Fellner, P., 311[42], 312[28], <u>314</u>, <u>315</u>
Fesseneden, J.M., 177[2], <u>198</u>
Fitz-James, P.C., 68[35], <u>86</u>
Flaks, J.G., 16[37], <u>61</u>
Flessel, C.P., 49[87], 57[87], <u>65</u>

Folk, W.R., 215[14], <u>230</u>
Forget, B., 288[12], <u>307</u> [13], 311[44], <u>313</u>, <u>314</u> <u>315</u>
Fraenkel-Conrat, H., 80[57], <u>88</u>, 273[12], <u>277</u>
Franklin, R.M., 48[71,72], 49[72], 57[72], <u>64</u>, 286[7], <u>313</u>
French, C.S., 184[22], <u>199</u>, 222[48], <u>233</u>
Friedman, S.M., 68[12], <u>85</u>

G

Galasinski, W., 178[16], <u>199</u>
Gallo, R.C., 260[33], <u>265</u>
Ganoza, M.C., 15[31], 32[31], <u>60</u>
Gardner, R.S., 31[55], 32[55,57], <u>62</u>, <u>63</u>
Garen, A., 3[9], <u>59</u>
Garner, C.W., 178[10], <u>198</u>
Gelfand, D.H., 113[15,17,20] 117[15,17,20,22], <u>119</u>, <u>120</u>
Genuth, S.M., 236[3], <u>263</u>
Gesteland, R.F., 2[6], 5[16], 19[6], 23[6], 29[6], 30[52], 45[52], <u>59</u>, <u>62</u>, 127[6], <u>148</u>, 273[15], <u>277</u> 281, <u>312</u>
Gibbons, N.E., 91[9,11], 92[10], 93[11], <u>109</u>, <u>110</u>
Gierer, A., 238, <u>263</u>, 273[14] <u>277</u>
Gilbert, W., 15[29], 16[33], 31[29,33], <u>60</u>, <u>61</u>, 268[6], <u>276</u>
Gilham, P.T., 380[40], <u>315</u>
Gillam, T., 204, <u>212</u>, 261[36], <u>265</u>
Goehnauer, M.B., 91[12], 93[12], <u>110</u>
Godson, G.N., 41[52], 49[85,86], 53[86], 54[85,86], 55[85,86], 56[85,86] 57[84,85], <u>63</u>, <u>65</u>, <u>125</u>, <u>148</u>, 184[21], <u>199</u>, 283, <u>312</u>

Goehler, B., 68[20], <u>85</u>
Gold, L.M., 29[50], <u>62</u>, 113[8,9,10,11], <u>119</u>
Goldemberg, S.H., 68[13], <u>85</u>
Goldstein, J., 211[13], <u>212</u>, 236[6], 238[12], 250[12], 251[23], 257[26], 258[26], <u>263</u>, <u>264</u>
Gonzalez, N.S., 48[76,79],49[76], 57[76], <u>64</u>, 68[13], <u>85</u>, 169[26], <u>176</u>
Gordon, J., 178[8], 179[19], 180[8], 181[20], 195[26], <u>198</u>, <u>199</u>
Gorini, L., 15[33], 30[53], 31[33,53], <u>61</u>, <u>62</u>
Gottlieb, D., 17[43,44],<u>61</u>
Gratzer, W.B., 291[152, <u>313</u>
Greenshpan, H., 161[8], 169[8], 172[8], <u>174</u>
Griffiths, E., 90[5,6,7], <u>109</u>
Gros, F., 14[24], 15[26], 17[24], 156[3], 160[15], <u>60</u> 151[3], 154[3], <u>174</u>,<u>175</u>,<u>216</u>[30], 225[30], 226[30], 229[30], <u>232</u>
Grossman, L., 49[86], 54[86],55[86], 56[86], 57[86], <u>65</u>, 113[1], <u>118</u>, 122[3], 132[12], 138[3], <u>147</u>, 186[23], 187[23], <u>199</u>, 273[13], <u>277</u>, 283[4], 288, 307[32], 308[34], <u>312</u>, <u>313</u>, <u>314</u>
Grubman, M., 146[16], <u>148</u>
Gulyas, A., 274[16], <u>277</u>
Gunsalus, I.C. 89[1], <u>109</u>
Guthrie, G., 48[70], <u>63</u>, 146[7] <u>149</u>

H

Hachmann, J., 192[25], <u>199</u>
Haines, J.A., 215[11], <u>230</u>
Hall, R., 307[32], <u>314</u>
Halvorson, H.O., 68[6], 68[38], <u>84</u>, <u>87</u>
Hänggi, J,U., 308[37], <u>315</u>
Hardesty, B., 178[15], <u>198</u>
Harewood, K., 251[23], <u>264</u>
Harris, J.I., 80[57], <u>88</u>

Hartley, B.S., 216[21,22],
  225[22], 226[21,22],231
Hayashi, H., 113[18], 117
  [18], 120, 215[17],
  221[17], 222[17], 225
  [17], 227[17], 229[17],
  231
Hayashi, M., 113[14,15,16,17
  18,20], 117[14,15,16,17,
  18,20,22], 119,120
Hayashi, M.N., 113[18],
  117[18], 120
Heinemeyer, C., 220[38], 232
Heinrikson, R.L., 216[21],
  226[21], 231
Herbert, E., 147[18], 148
Hershey, J.W.B., 161[16],
  162[117], 172[17],157
Herzberg, M., 156[11], 164
  [11], 161[18],169[18],
  172[8], 174
Heywood, S..M., 154[6], 174
Hiatt, H.H., 15[30], 60
Hill, W.E., 285[6], 313
Hille, M.B., 15[28],60,
  170[20], 171[28], 175,
  176
Hirashima, A., 68[3], 84
Hirs, C.H.W., 219[36], 232
Hirsh, D,I. 216[28], 221
  [28,43], 225[28], 228
  [43], 231, 232
Hirshfield, I.N., 222[47],
  233
Hoagland, M.B., 217[32],
  320[32], 232 236,
  236[2], 237, 263, 264
Holdsworth, E.S., 68[10],
  85
Holl, H., 285[5], 312
Hollingworth, B.R.,41[61],
  45[61], 63
Holley, R.W., 236[6], 238
  [13, 17], 257[27,28,29],
  263, 264
Horikoshi, K., 68[50], 87
Hotham-Iglewski, B., 48
  [71,72], 49[72], 57[72],
  64, 286[7], 313

Howard, G., 147[18], 148
Hsieh, W.T., 268[6], 276
Hultin, T., 236[7], 263

I

Idriss,J.M., 68[38], 87
Igarashi, R.T., 68[19,30], 85
  86
Ionesco, H., 70[52], 88
Imsande, J., 68[7], 84
Isham, K.R., 68[22],85
Itoh, T., 68[40], 87
Iwasaki, K., 162[18], 167[7],
  168[18,24], 169[24], 137
  [24], 174[18,24], 176

J

Jacob, F., 128[8], 148
Jacobson, K.B., 222[46], 233
Jones, D.S., 32[60], 63
Jones, O.W., 32[58], 63
Jordan, B., 311[44], 315
Joseph, D.R., 216[29], 225[29],
  229[29], 231
Julian, G.R., 31[56], 62

K

Kaempher, R., 48[67,68,82,81],
  55[68,83], 57[68], 63,64,
  122[1], 122[2,3], 128[1],
  132[12], 135[1,2], 138[32],
  145[14], 146[14,15], 147,
  148
Kaneko, I., 68[18,19,20], 85
Kaplan, N.D., 6[15], 9[15],59
  75[55], 77[55], 88, 106[18],
  110, 184[22[, 199, 222[48],
  233
Kaplan, S., 280[1], 312
Katchalski, E., 31[54], 62
Kati, T., 48[80], 64
Katze, J.R., 215[17], 216[27],
  221[17,27,45[, 222[17,27],
  225[17,27], 227[17], 229
  [17,27], 231, 233
Keller, E.B., 236[2], 262

Kelmers, A.D., 259, 261[39], 265
Khorana, H.G.,32[60], 63
Kiho, Y., 58[89], 65
Kindley, J., 297[22], 299 [22], 314
Kirby, K.S., 106[16], 110, 238, 246, 250, 263, 264, 288, 313
Klein, H.A., 211[14], 212
Knowles, J.R., 215[17], 221 [17,45], 222[17], 225[17], 229[17], 231, 233
Kobayashi, Y., 68[6], 84
Kohler, R.E., 48[74,75], 49 [84], 50[84], 53[84], 54[84], 57[84], 58[74], 65
Kolakofsky, D., 162[17], 165 [22], 172[17], 175
Kondo, M.N,68[43], 87
Konigsberg, W.,6[14], 12[20], 16[14], 21[14,20,47], 22 [14], 28[14], 46[20], 48[-20], 59,60,62, 216[27], 221[27,45], 222[27], 225 [27], 229[27], 231,233
Koprowki, A., 30[51], 62
Kornberg, A., 68[8,39], 84, 87, 218[35], 232
Kosakowski, M.H.J.E., 216[25] 226[25], 229[25], 231
Krisko, I., 179[9], 199
Kull, F.J., 222[46], 233
Kushner, D.J., 89[1], 91[12], 93[12], 98[15], 109,110

L

Lamborg, M.R., 300[23], 314
Lapointe, J., 215[17], 221 [17], 222[17], 225[17], 227[17], 229[17], 231
Larsen, H., 89[1,2], 109
Laskin, A.I., 17[44], 61
Last, J.A., 15[28], 60, 170 [20], 175
Lazzarini, R.A.,68[23,24], 86, 215[13], 225[13], 226[13], 230
Leder, P., 32[58], 63, 80[56], 88, 171[21], 176, 178[9], 198, 204[5], 206, 212, 256 [25], 264,
Lederberg, S., 68[45], 87
Lederberg, V., 68[45], 87
Lederman, M. 113[2,4,5,6,], 118, 119
Lee, J.C., 308[40], 315
Lee, L.W., 220[39], 232
Lee.,M., 216[26], 221[26], 222 [26], 225[26], 231
LeLong, J.C., 156[11], 160[15], 164[11], 175
Lemoine, F., 216[23], 225[23], 226[23], 227[23], 229[23], 231
Lengyel, P., 10[17], 16[38], 31[55], 32[55,57,59], 59 61,62,63, 68[11,26,27,28, 29], 85, 86, 178[13], 179 [17], 198, 199, 214[4],230
Lennox, E.S., 4[10],59
Levinthal, C., 295[18], 297 [19,20], 313, 314
Levy, A.L., 80[57], 88
Lipmann, F., 10[16], 18[16], 59, 68[5], 71[53], 84, 88, 106[17], 110, 177[4], 178 [5,6,14], 179[18,19], 198, 199, 204, 212, 221[43], 228 [43], 232
Littauer, U.Z., 261[35], 265
Lodish, H.F., 2[5], 5[11], 6[11], 8[11], 15[5], 16 [35], 59, 64, 154[4], 174
Loeb, T., 268[5], 276
Loening, U.E., 251[24], 264,292 [16], 313
Loewus, F.A., 246[20], 264
Loftfield, R.B., 214[5], 221 [42], 230, 232
Lowenstein, J.M., 114[21], 120
Lowry, O.H., 98[14], 110, 246 [19], 264
Lucas-Lenard, J., 178[5], 198, 204[5], 212

Author Index

Lenard, J., 179[18], <u>199</u>
Lundgren, D.G., 68[34], <u>86</u>
Luzzatto, L., 13[22], 16[22], 32[22], <u>60</u>

M

Maden, B.E.H., 18[45], <u>61</u>
Madison, J.T., 257[29], <u>264</u>
Magasanik, B., 303[31], <u>314</u>
Main, R.K., 228[50], <u>233</u>
Maitra, U., 163[19], 167[19], 169[19], <u>175</u>
Mangiarotti, G., 48[66,69], 49[66], 58[66], <u>63</u>
Maramorosch, K., 30[51], <u>62</u>
Markham, R. 303[29], <u>314</u>
Marmur, J., 68[32], <u>86</u>, 118 [23], <u>120</u>
Marotta, C.A., 308[38], <u>315</u>
Marquisee, M., 257[29], <u>265</u>
Marrs, B., 280[1], <u>312</u>
Marshal, R.D., 228[52], <u>233</u>
Martin, R.G., 291[14], <u>313</u>
Matthaei, J.H., 2[1], 6[1], 8[1], 9[1], 10[1], 14 [25], 22[1], <u>58</u>,<u>60</u>
Maxwell, O.H., 262[37], <u>265</u>
Mazumder, R., 156[12], 158[12] 162[18], 167[12], 168[12] 173[18], <u>175</u>

Mc

McCance, M.E., 91[11], 93[11] <u>110</u>
McConkey, E., 214[5], <u>230</u>
McCorquodale, D.J., 229[53], <u>233</u>
McElroy, W.D., 221[44], <u>232</u>
McGrath, J., 113[19], <u>120</u>
McKeehan, W.L., 178[15], <u>198</u>
Mehler, A.H. 215[10,13,19], 220[40], 221[44], 225[10, 13,19], 226[10,13,19], <u>230</u> <u>231</u>, <u>232</u>
Merrill, S.H., 238[17], 257 [28,29], <u>264</u>, <u>265</u>
Meselson, M., 48[67,82], <u>63</u>, <u>64</u>, 122[1], 124, 128[1,9], 132[12], 135[1], 145[14], 146[14], <u>147</u>, <u>148</u>
Miall, S.H., 48[80], <u>64</u>
Migita, L.K., 68[1,2,50], <u>84</u> <u>87</u>
Miller, D.L., 192[25], <u>199</u>
Miller, M.J., 162[18], 168[18], 174[18], <u>175</u>
Miller, R.S., 31[55], 32[55,57] <u>62</u>
Millward, S., 204[7], <u>212</u>, 261 [36], <u>265</u>
Milman, G., 211[13],
Milner, H.W., 184[22], <u>199</u>,222 [48], <u>233</u>
Mingioli, E.S., 49[88], 52[88], <u>65</u>
Mitra, S.K. 215[10], 220[40,41] 225[10], 226[10], <u>230</u>,<u>232</u>
Modolell, J., 13[23], 14[23], 16[23], 26[23], 35[23], 42 [23,63,64], 43[23], 44 [23,63], 46[23], 47[23], <u>60</u> <u>63</u>
Moldave, K., 49[86], 54[86], 55 [86], 56[86], 57[86], <u>65</u>, 106[18], <u>110</u>, 113[1], <u>118</u>, 132[12], <u>138</u>[3], <u>148</u>,177[2] 178[16], 186[23], 187[23], <u>198</u>, <u>199</u>, 273[13], <u>277</u>,283 [4], 288, 307[33], 308[34], <u>312</u>, <u>313</u>, <u>314</u>
Mouro, R.E., 18[45], <u>61</u>, 177[4] <u>198</u>
Moon, H.M., 68[26], <u>86</u>, 178[13] <u>198</u>
Moore, S., 219[36], <u>232</u>
Morell, P., 68[32], <u>86</u>
Muench, K.H., 205, <u>212</u>. 215 [9,12], 216[26,29], 221[9, 12,26], 225[26,29], 226 [12], 227[2], 228[12,51], 229[29,54], <u>230</u>,<u>231</u>, <u>233</u>, 259, 260[32], <u>265</u>

N

Nakada, D., 146[16], <u>148</u>

Nakamoto, T. 15[31], 32[31] 60
Nanninga, N., 68[41], 87
Nathans, D., 2[2,3], 6[2,3], 8[2,3], 12[2,3], 15[2], 21[2], 21[3], 58, 59
Nau, M.M., 178[9], 198
Neidhardt, F.C., 214[8], 230
Newton, G., 236[1], 263
Nirenberg, M.W., 2[1], 6[1, 15], 8[1], 9[15,1], 10[1], 14[25], 22[1], 32[58], 58, 59, 60, 63, 97[13], 110, 171[21], 176, 202[2], 203[3], 205[9], 206, 211, 212, 256[25], 264
Nishihora, T., 272[11], 277
Nishizuka, Y., 178[6], 198
Nisman, B. 113[1], 118
Noll, H. 130[-1], 131[11], 132[12], 148
Nomura, M., 48[70], 63, 128[9], 146[17], 148, 149
Norris, J.R., 91[9], 109
Nossal, N.G., 225[49], 233
Notani, G., 2[2,], 6[2], 8[2], 12[2], 15[2], 21[2,20], 46[20], 48[20], 58, 60
Novelli, G.D., G.214], 230 236[3], 237[10], 263

O

Ochoa, S., 15[28], 32[59], 60,61,63, 151[1], 154[1], 156[12], 157[13], 158[12], 167[7], 168@24], 169[24], 170[20], 171[28], 173[24], 174[24], 175, 176
Ohta, T.,160[14], 165[14], 175
Oleinick, N.L., 17[42], 61 68[47], 87
Onishi, H., 90[11], 93[11], 110
Ono, Y., 68[26,27,28,29],
86, 178[13], 198
Osawa, S., 68[40], 87
Ostrem, D.L., 215[15], 231
Otaka, E., 68[40], 87

P

Pace, B., 311[43], 315
Pace, N.R., 311[43], 315
Parmegglani, A., 178[7], 198
Peacock, A.C., 292[17], 298, 299[17], 313
Pearson, R.L., 261[39], 265
Peng, C.H.L., 218[34], 232
Penswick, J.R., 257[29], 265
Pestka, S., 32[58], 63 260[33], 265
Peterson, P.J., 214[3], 230
Peterson, R.L., 311[43], 315
Pfister, R.M., 68[34], 86
Phillips, L.A.,48[71,72], 49[72] 57[72], 64, 286[7], 313
Price, C.A., 288[10], 313
Pricer Jr., W.E., 75[55], 77[55], 78, 218[35], 232

R

Rabinowitz, J.C., 75[55], 77[55], 88
RajBahandry, U.L., 308[39], 315
Ralph, P., 49[87], 57[87], 65
Randall, R.J., 98[14], 110, 264[19], 264
Raskas, H.J., 48[67], 63, 122[1], 128[1], 135[1], 147
Ravel, J.M., 178[10,11], 198, 220[38,39], 232
Redfield, B., 178[12], 198
Reeder, R.H., 295[18], 313
Remold-O'Donnell, E., 162[17], 166[23], 171[23], 172[17], 175
Ravel, M., 15[26,30], 60,156[11], 160[15], 161[18],164 109[8], 172[8], 174[18], 175
Reynier, M., 300[24], 314
Ribbons, D.W., 91[15], 109

Rich, A., 49[87], 57[87], 58
    [89], 65
Richards, E.G., 291[15], 313
Richter, D., 178[14], 198
Richenberg, H.V., 281, 312
Roberts, R.B., 12[19], 60,
    129, 130[10], 148
Robertson, H.D., 2[5], 15
    [5], 59
Roblin, R., 267[3], 276
Ron, E.Z., 48[74, 75, 77],49
    [84], 50[84], 53[84], 54
    [84], 57[84], 58[74], 64
    65
Rosebrough, N.J., 98[14],
    110, 246[19], 264
Rosen, 290[13], 313
Rossetti, G.P., 285[6], 313
Rouget, P., 215[18], 218[18]
    231
Ryter, A., 70[52], 88

S

Sabol, S., 162[18], 167[7],
    168[18,24], 169[24], 173
    [24], 174[18,24], 175,
    176
Sacks, LE, 73[54], 88
Safille, P.A., 215[12], 222
    [12], 226[12], 227[12],
    228[12], 230, 260[32],
    265
Salas, M., 15[28], 60, 151[1],
    154[1], 157[13], 170[20],
    171[28], 174, 175, 176
Salser, W., 2[6], 15[6], 19[6]
    23[6], 29[6], 59
Sanger, F., 303[26], 308[26],
    [36], 311[41], 314, 315
Santangelo, E., 68[24], 86
Sargent, J., 285[5], 312
Saunders, G.F., 68[44], 87
Schaeffer, P., 70[52], 88
Schechter, N., 68[15], 85
Schimmel, P.R., 220[37], 232
Schleich, T., 238[12], 250[12]
    263
Schlessinger, D., 13[22], 14

[24], 16[22], 17[24], 32
    [22], 41[61], 45[61], 48
    [66, 69], 49[66], 58[66],
    60, 63, 68[9], 85, 128[9],
    148, 287, 313
Schofield, P., 269[40], 265
Schramm, G., 238, 263, 273[14],
    277
Schwartz,J.H., 2[2], 6[2,12],
    8[2,12], 12[2], 15[2], 21
    [2], 58, 59
Schweet, R.S., 177[1,3], 197,
    198
Schweiger, M., 29[50], 62, 113
    [8,9,10,11], 118
Scolnick, E., 202[2], 203, 211
    [11, 12, 13],
Sehgal, S.W. 92[10], 109
Sekiguchi, M., 29[49], 62
Sela, M., 31[54], 62
Shafritz, D.A., 154[5], 174
Shaw, P.D., 17[43,44], 61
Shive, W., 178[10], 198, 220
    [38,39], 232
Shorey, R.L., 178[10], 198
Siddigi, O., 3[9], 59
Sih, C.J., 68[46], 87
Sillero, M.G., 162[18], 168
    [18,24], 169[24], 173[24],
    174[18,24], 175, 176
Singer, M., 204[6], 212, 225
    [49], 233
Sinsheimer, R.L., 49[85], 54
    [85], 55[85], 56[85], 57
    [85], 65, 125, 148, 184[21]
    199, 267[3], 270, 273[9],
    276[9]
Skogerson, LE, 178[9], 198
Skoultchi, A., 68[26,27,28,29],
    86, 178[13], 198
Slater, D.W., 3[7], 6[13], 59
Sly, W.S. 32[58], 63
Smith, A.M., 157[13], 175
Smith, C.J., 220[41], 232
Smith, I., 68[32], 86
Smith, J.D., 303[29], 308[34,
    [35], 314
Smith, M.A., 18[46], 31[46], 62
Söll, D., 10[17], 59, 179[17],
    199, 214[4], 215[17], 221
    [17], 222[17], 225[17],

227[17], 229[17], <u>230</u>, <u>231</u>, <u>233</u>
Solymosy, F., 274[16], <u>277</u>
Speyer, J.G., 16[38], 31[55], 35[55,57,59], <u>61</u>, <u>62</u>, <u>63</u>
Spiegelman, S., 3[7], 6[13], <u>59</u>
Spirin, A.S., 301, <u>314</u>
Staehlin, T., 147[19], <u>148</u>, 244, <u>264</u>, 287[8], <u>313</u>
Stahman, M.A., 18[46], 31[46], <u>62</u>
Stanier, R.Y., 89[1], <u>109</u>
Stanley, W..M., Jr., 151[1] 154[1], 157[13], 171[28], <u>174</u>, <u>175</u>, <u>176</u>, 274[17], <u>277</u>, 303[27], <u>314</u>
Stap, F., 48[81], <u>64</u>
Stein, W.H., 219[36], <u>232</u>
Stenesh, J., 68[15], <u>85</u>
Stephenson, M.L., 237[9], 238[14], <u>263</u>
Stern, R., 215[19], 221[44] 225[19], 226[19], <u>231</u>, <u>232</u>, 261[36], <u>265</u>
Stevens, A., 68[22], <u>85</u>
Stevens, L., 68[42], <u>87</u>
Strauss, J.H., Jr., 267[3], 270, 273[9], 276[9], <u>276</u>, <u>277</u>
Streeck, R.E., 308[37], <u>315</u>
Stulberg, M.P., 68[22], <u>85</u> 216[24], 221[24], 225[24] 226[24], 229[24], <u>231</u>
Subramanian, A.R., 48[77, 78], <u>64</u>, 169[25], 174[23], <u>176</u>
Sueoka, N., 68[21], <u>85</u>, 261[34], <u>265</u>
Sugiura, M.,113[14], 117[14] <u>119</u>
Szekely, M., 308[36], <u>314</u>
Szer, W., 15[32], <u>61</u>

T

Takai, M., 68[43], <u>87</u>
Takeda, M., 68[4,5], <u>84</u>

Tamaoki, T., 48[80], <u>64</u>
Tener, G.M., 204[7], <u>212</u>, 261[36], <u>265</u>
Thach, R.E., 160[14], 161[16], 162[17], 165[14],22], 166[23], 171[23], 172[17], <u>175</u>
Thompkins, R., 202[2], 203[3], 211[12], <u>212</u>
Tishler, P.V., 297[21], <u>314</u>
Tissieres, A., 14[24], 17[24], 41[51], 45[61], <u>60</u>, <u>63</u>, 113[12], <u>119</u>
Traut, R.R., 18[45], <u>61</u>, 288[10], <u>313</u>
Tsugita, A., 68[3], <u>84</u>

V

Van Holde, K.E., 285[6], <u>313</u>
Van Knippenberg, P.H., 48[81], <u>64</u>
Van Rapenbusch, R., 216[23], 225[23], 226[23], 227[23], 229[23], <u>231</u>
Varricchio, F., 307[33], <u>314</u>
Vazquez, D., 17[41,43], <u>61</u>
Vischer, E., 303[31], <u>314</u>
Voigt, H.P., 308[37], <u>315</u>
Vold, B., 68[25], <u>86</u>
von Ehrenstein, G., 71[53], <u>88</u>, 106[17], <u>110</u>
von Tigerstrom, M., 204[7], <u>212</u> 261[36], <u>265</u>
Voorma, H.O., 48[81], <u>64</u>

W

Wade, H.E., 3[8], <u>59</u>
Wahba, A.J., 15[28], 31[55], 33[55, 57, 59], <u>60</u>, <u>62</u>, <u>63</u>, 151[1], 154[1], 157[13], 162[18], 167[7], 168[18], 170[20], 171[28], 174[8], <u>174</u>, <u>175</u>, <u>176</u>
Waldenstrom, J., 215[20], 225[20], 226[20], 229[20], <u>231</u>
Waller, J.P., 216[23], 225[23], 226[23], 227[23], 229[23], <u>231</u>

Author Index

Waltho, J.A, 90[4], <u>109</u>
Wang, S., 220[38], <u>232</u>
Waskell, L., 113[19], <u>120</u>
Watanabe, I., 272[11], <u>277</u>
Waterson, J., 68[11,27, 28, 29], <u>85</u>, <u>86</u>
Watson, J.D., 41[61], 45[61], <u>63</u>
Weber, K., 21[20, 47], 46[20], 48[20], <u>60</u>, <u>62</u>
Webster, R.E., 6[4], 16[14, 34a], 21[14], 22[14], 28[14], 41[34a], 42[34a], 44[34a], 46[65], 47[34a], <u>59</u>, <u>61</u>, <u>63</u>, 68[4], <u>84</u>
Weigle, J., 124, <u>148</u>
Weinstein, I.B., 68[12], <u>85</u>
Weisblum, B., 17[40], 42[40], 44[40], <u>61</u>, 68[46], <u>87</u>, 257[27], <u>264</u>
Weiss, J.F., 259, <u>265</u>
Weissbach, H., 178[12], 192[25], <u>198</u>, <u>199</u>
Weissman, C., 268[7a], 270[7a,] <u>276</u>
Weith, H.L., 308[40], <u>315</u>
White, J.R., 16[37], <u>61</u>

Wilhelm, J.M., 17[42], <u>61</u>, 68[47], <u>87</u>
Wilkins, M.J., 228[50], <u>233</u>
Williams, L.S., 214[8], <u>230</u>
Wimmer, E., 204[7], <u>212</u>, 261[36], <u>265</u>
Wyatt, G.R., 303[30], <u>314</u>
Woese, C., 68[33], <u>86</u>

Y, Z

Yamane, T., 261[34], <u>265</u>
Yaniv, M., 216[30], 225[30], 226[30], 229[30], <u>232</u>
Young, E.T., 113[12,13], <u>119</u>
Zachau, H.G., 308[37], <u>315</u>
Zamecnik, P.C., 215[11], 228[52], <u>230</u>, <u>233</u>, 236[2], 237[9], 238[14], <u>262</u>, <u>263</u>
Zamir, A., 257[29], <u>265</u>
Zinder, N.D., 2[2], 6[2,14], 8[2], 12[2], 15[2,20], 16[14, 34a], 21[2,14,20], 22[14], 28[14], 41[34a], 42[34a], 44[34a], 46[20], 46[65], 47[34a], 48[20], <u>58</u>,<u>59</u>,<u>60</u>,<u>61</u>,<u>63</u>, 268[5], <u>276</u>
Zubay, G., 113[2, 3, 4, 5, 6, 7], <u>118</u>, <u>119</u>

SUBJECT INDEX

A

Amino acid incorporation; see Polypeptide synthesis
Aminoacyl-tRNA binding factors, 178, 179; see also Elongation factors
Aminoacyl-tRNA formation, 107-109, 218-220
Aminoacyl-tRNA synthetases, 104, 213 ff.
 assay, 107-109, 216-221
 preparation, 104-105
 purification, 221-229
 storage, 229
Amino-terminal residues, 80-81
Antibiotics, 16-17; see also individual antibiotics
ATP-PP$_i$ exchange reaction, 217-218

B

Bacillus subtilis, 67 ff.
Bacteriophage; see Phage
BD-cellulose, for fractionation of tRNA, 261
Bentonite, 30, 73, 290

C

Cell disruption, 6
 Bacillus subtilis cells, 75
 Bacillus subtilis spores, 76
 blending with glass beads, 184, 223
 EDTA-lysozyme, 48, 54-57, 125, 283
 Escherichia coli; 6, 115, 125, 155, 183, 223, 249, 269, 283
 freeze-thaw-lysozyme, 48, 49-54

(Cell disruption, cont'd.)
 French pressure cell, 6, 9, 115, 184, 223, 283
 grinding with alumina, 6-7, 155, 183, 249, 283
 Halobacterium, 96-98
 hydrolytic enzymes, 6
 Halobacterium, 96-98
 hydrolytic enzymes, 6
 lysis, 58, 96, 125, 269
 Manton-Gaulin mill, 184
 for polysomes, 48-58, 125
 for ribosomal RNA, 282-283
 sonication, 6, 184, 223, 283
Cell growth, 4
 Bacillus subtilis, 70-71
 Escherichia coli, 4, 114, 124, 155, 181-182, 203, 222, 268, 281
 Halobacterium, 91-94
 for labeling cells, 124
 for phage preparation, 268-270
 for ribosomal RNA, 281-282
Cell harvesting, 5, 70, 94-96, 114, 155, 182-183, 222
Cell storage, 5-6, 70-71, 114, 155, 183, 222
Cesium chloride gradients, 128
 preparation, 129-133
Chain elongation factors; see Elongation factors
Chain initiation factors; see Initiation factors
Chain termination factors; see Termination factors
Chloramphenicol, 14, 16, 35, 38, 42-43, 51-52, 57, 127
Codons, 204
Coliphage; see Phage
Countercurrent distribution, for fractionation of tRNA, 257-259

D

DEAE-cellulose, 244
Density gradients, 129 ff.; see also Sucrose density gradients
  cesium chloride, 128-133, 137-144
  experimental, 131
  glycerol, 129-133
  linear, 129-131
  preparation, 33-34, 129-133
  resolving power, 133
  sucrose, 36-40, 129-133, 133-137
Dialysis, 8
  tubing, cleaning of, 10
Dithiothreitol (DTT, Cleland's reagent), 12, 20, 180
DNA-dependent RNA polymerase, 111, 114
DNA templates, 117
  calf thymus DNA, 117
  Escherichia coli DNA, 118
  λ phage DNA, 118
  φX-174 RF DNA, 117
  SV40 DNA, 117
  T2 DNA, 117
  T4 DNA, 117
  T7 DNA, 117

E

Elongation factors (G, T, T_u, T_s), 177 ff.
  assays, 194-197
  elongation cycle, 18
  preparation, 179-188
  purification of T and G factors, 189-191
  separation of T and G factors, 188-189
  separation of T_u and T_s, 192
  terminology, 178-179
Erythromycin, 16
Escherichia coli strains
  B, 3, 53, 155, 203, 202
  C, 114, 182
  D10(IU, D1IU), 114, 281
  Hfr 3000, 268
  K12, 3, 53, 155, 182, 222

(Escherichia coli, cont'd.)
  MRE 600, 3-4, 155, 281
  Q13, 3, 114, 155, 268
  S/6, 182
  S26, 3-4
  W, 53
Escherichia coli strains
  deficient in polynucleotide phosphorylase, 3
  deficient in RNase, 3, 155, 268, 281

F

f Factors; see Initiation factors
Factors R1 and R2; see Termination factors
"Finger nuclease", 25, 240
N-formylmethionine, 77-78
  isolation of, 79-80
N-formylmethionylalanine, 77
  isolation of, 80
N-formylmethionyl-tRNA (fMet-tRNA), 3, 15
  binding to ribosomes, 170-171
  f[^{3}H]methionine release assay, 203 ff.
  f[^{3}H]Met-tRNA^f, 204
  f[^{3}H]Met-tRNA-AUG-ribosome complex, 206-207
Formyltetrahydrofolic acid synthetase, 74-75

G

Gel filtration for purification of tRNA, 251-256
GDP binding assay for T factor, 195
G factor; see Elongation factors
Glutathione, 20, 180
Glycerol gradients, 129-138
  resolving power of, 133-137
GTPase activity, 172
GTPase assay for G factor, 196-197

Subject Index

H

Halocacterium, 89 ff.
 aminoacyl-tRNA synthetases, 104
 cell disruption, 96-98
 cell growth, 91-94
 cell harvesting, 94-96
 media, 91-92
 ribosomes, 101
 S-60 system, 96
 S-150 system, 101
 tRNA, 104
Halophilic bacteria, 89 ff.
Homopolynucleotides; see synthetic polynucleotides
Hydroxamate assay, 220-221

I

Initiation factors (f Factors, f1, f2, f3), 151 ff.
 assays, 169-174
 crude, preparation of, 11, 154-159
 f1, assays for, 170-171
 f1, preparation of, 159-163
 f2, assays for, 171-172
 f2, preparation of, 163-167
 f3, assays for, 172-174
 f3, perparation of, 167-169
 sources, 154
 terminology, 152-153

M

MAK; see Methylated albumin-kieselguhr
Methylated albumin-kieselguhr (MAK), 261
 fractionation of tRNA, 261
 rRNA, 300
Methylated albumin-silicic acid (MASA), 261

P

Partition chromatography for fractionation of tRNA, 259-260

Partition systems, 186-188, 257-260
Penicillin, 58
Phage (Bacteriophage), 267 ff.
 growth, 268
 isotopically labeled, 270
 purification, 270-273
 RNA, 267 ff.
 RNA as messenger, 14-15, 18, 37, 118, 267
Phage RNA, 267 ff.
 isotopically labeled, 270
 preparation, 273-276
 storage, 276
Phage RNA as messenger, 14-15, 18, 267
 f2, 14-15
 MS2, 14-15, 118
 Qβ, 14-15, 118
 R17, 4, 37
Phenol extraction; 238 ff.
 Bacillus subtilis tRNA, 71-73
 Halobacterium tRNA, 106
 phage, 273-276
 ribosomal RNA, 288-290
Polyacrylamide gel for rRNA separation, 291-300
Polypeptide chain length, estimation of, 47-48, 81
Polypeptide synthesis, 12 ff.
 analysis by sucrose density gradients, 32 ff.
 in Bacillus subtilis systems, 76-84
 directed by endogenous mRNA, 12, 29, 76-81
 directed by phage RNA, 14-15 18 ff., 37, 83, 118
 directed by poly U, 14, 30, 81-83, 118
 directed by synthetic polynucleotides, 14, 30-32, 81-83, 103-104
 directed by T4 messenger, 14, 15, 28-29
 in Escherichia coli systems, 12 ff.
 estimation of rate, 46-47

(Polypeptide synthesis, cont'd.)
  in Halobacterium systems, 99–101
  inhibition by streptomycin, 16
Polypeptide termination factors, see Termination factors
Polynucleotides; see Synthetic polynucleotides
Polysomes, 48 ff.
  concentration, in subunit exchange, 127–128
  formation in vitro, 17
  from fragile cells, 58
  preparation, 48–58, 124–126
  from spheroplasts, 57
  in sucrose gradient profiles, 41, 51
Poly U, 14, 18, 30, 81–83, 118, 194–195
Poly U system for elongation factors, 194–195
Protamine sulfate, 186
Puromycin, 14, 18, 39, 44
Pyrophosphate exchange reaction; see ATP-PP_i exchange reaction

R

R1 and R2 factors; see Termination factors
Release assay, 207–208
Ribosomal RNA, 279 ff.
  polyacrylamide gel fractionation, 291–300
  sucrose gradient fractionation, 290–291
Ribosomal subunit exchange, 122 ff.
  applications, 144–146
  principle, 122–123
  reaction, 126
  relative rates, 144–145
Ribosomal subunits, 121 ff.
  isotopically heavy, 124
  isotopically light, 124
  separation of, 286–287
  in sucrose gradients, 40

Ribosomes, 10–12
  binding to phage DNA, 172
  Escherichia coli, 10–12, 115–116, 156, 192–194, 203–204, 284
  estimation of number active, 43–45
  Halobacterium, 101–103
  labeled, 125
  preparation, 10–12, 101–103, 115–116, 156, 192–194, 284
  sucrose gradients, 40, 285–286
  washing, 157, 203–204
RNA: see Ribosomal RNA, tRNA, Phage RNA, 5S RNA
RNA polymerase; see DNA-dependent RNA polymerase
RNA synthesis, 116–117
tRNA; see Ribosomal RNA
5S RNA, 291, 294, 300

S

S-30 extracts, 2
  assay 2–6
  Bacillus subtilis, 75
  Escherichia coli, 6–10, 156
  preparation, 6–10
  storage, 9
S-60 extracts, from Halobacterium, 96 ff.
  assay, 98–101
  preparation, 96–98
  storage, 98
S-100 extracts, 10–12
  preparation, 10–12, 156, 248
S-150 extracts, from Halobacterium, 101
Sparsomycin, 126, 127
Spectinomycin, 16
Spheroplasts, 283–284
Spores, Bacillus subtilis, 73–74
Streptogramin A, 17
Streptomycin, 14, 16, 38, 42–43, 47
Subunits; see Ribosomal subunits

Subject Index

Sucrose density gradients, 36-40
 analysis of polypeptide synthesis, 32 ff.
 analysis of polysomes, 51
 interpretation, 40-48
 preparation, 33-34
 resolving power, 133
 for ribosomes, 40, 285-286
 for rRNA, 290-291
 for S values, 291
S value, of rRNA, 291, 300
Synthetic mRNA; see Synthetic polynucleotides
Synthetic polynucleotides (homopolynucleotides, polyribonucleotides, synthetic mRNA), 14, 17, 30-32, 103-104

T

Termination factors, 201 ff.
 assays, 207, 211
 preparation of R1 and R2, 208-210
 purification of R1, 210
 purification of R2, 210-211
 release assay, 207-208
Tetracycline, 17
T factor; see Elongation factors
T_s factor; see Elongation factors
T_u factor; see Elongation factors
Translocase, 179
tRNA, 235 ff.
 Bacillus subtilis, 71-73
 Escherichia coli, 235 ff.
 N-formylmethionyl-tRNA, 170-171
 fractionation, 257-262
 Halobacterium, 104-107
 isolation, from disrupted cells, 248-250
 isolation, from whole cells, 238-247
 nomenclature, 237-238
 purification, 250-257

U

Ultracentrifugal analysis, of ribosomal subunit exchange reaction, 128

Z

Zonal centrifugation, 35-36
 for ribosomal subunits, 288

09752
1/D2